T0330355

Platform Economy Puzzles

A Multidisciplinary Perspective on Gig Work

Edited by

Jeroen Meijerink

Assistant Professor of Human Resource Management, Faculty of Behavioural Management and Social Sciences, University of Twente, the Netherlands

Giedo Jansen

Assistant Professor of Public Administration, Faculty of Behavioural Management and Social Sciences, University of Twente, the Netherlands

Victoria Daskalova

Assistant Professor of Law, Governance and Technology, Faculty of Behavioural Management and Social Sciences, University of Twente, the Netherlands

 Edward Elgar
PUBLISHING

Cheltenham, UK • Northampton, MA, USA

Cover image: Oliver Cole on Unsplash.

Published by
Edward Elgar Publishing Limited
The Lypiatts
15 Lansdown Road
Cheltenham
Glos GL50 2JA
UK

Edward Elgar Publishing, Inc.
William Pratt House
9 Dewey Court
Northampton
Massachusetts 01060
USA

A catalogue record for this book
is available from the British Library

Library of Congress Control Number: 2021938809

This book is available electronically in the **Elgar**online
Political Science and Public Policy subject collection
http://dx.doi.org/10.4337/9781839100284

ISBN 978 1 83910 027 7 (cased)
ISBN 978 1 83910 028 4 (eBook)

Printed and bound by CPI Group (UK) Ltd, Croydon, CR0 4YY

Contents

Figures

Tables

Contributors

Adam Badger is a PhD candidate at the Schools of Geography and Management, Royal Holloway, University of London, United Kingdom.

Damion Jonathan Bunders is a PhD candidate at the Rotterdam School of Management, Erasmus University Rotterdam, the Netherlands.

Ronan Carbery is Senior Lecturer in Management and Co-Director of the Human Resource Research Centre at the Cork University Business School, University College Cork, Republic of Ireland.

Victoria Daskalova is Assistant Professor of Law, Governance and Technology at the University of Twente, the Netherlands.

James Duggan is Assistant Professor in Management at the School of Business, Maynooth University, Republic of Ireland.

Giedo Jansen is Assistant Professor of Public Administration at the University of Twente, the Netherlands.

Paul Jonker-Hoffrén is Senior Researcher at the Work Research Center of University of Tampere, Finland.

Anne Keegan is Professor of Human Resource Management at University College Dublin, Republic of Ireland.

Shae McCrystal is Professor of Labour Law at the University of Sydney Law School, Australia.

Anthony McDonnell is Full Professor of Human Resource Management and Co-Director of the Human Resource Research Centre at the Cork University Business School, University College Cork, Republic of Ireland.

Jeroen Meijerink is Assistant Professor of Human Resource Management at the University of Twente, the Netherlands.

Annarosa Pesole is Researcher at the Joint Research Centre – European Commission, Spain.

Aaron Shapiro is Assistant Professor of Technology Studies at the University of North Carolina at Chapel Hill, USA.

Ultan Sherman is Lecturer in Management and Organisational Behaviour at the Cork University Business School, University College Cork, Republic of Ireland.

Jim Stanford is Economist and Director of the Centre for Future Work, Australia and Canada.

Niels van Doorn is Assistant Professor in the Department of Media Studies, University of Amsterdam, the Netherlands.

Masako Wakui is Professor of Law at Kyoto University, Japan.

Preface and acknowledgements

This book is the result of our shared interest in the platform economy and a desire to conduct a joint research project at the interface of work, technology and society. Coming from different academic disciplines and working at a social science faculty within a technical university that revolves around 'High Tech, Human Touch', to us it was a logical choice to start this book project. In doing so, we aim to further a multidisciplinary approach to the study of the antecedents, outcomes and solutions to platform economy puzzles. Here, we want to thank those who supported us in making this book become a reality. First and foremost, we thank the contributors to this edited volume. It is the diversity in their expertise that made this book possible, and we thank them for their commitment and rich insights that their work offers. Second, we thank Koen Frenken, Charissa Freese, Christian Fieseler and Mike Maffie for their constructive comments on earlier versions of (selected chapters in) this edited volume, and Gemma Newlands for helping us to commit colleague-academics to our book project. Third, we thank the Smart Industry and Resilience themes of the Faculty of Behavioural, Management and Social Sciences of the University of Twente for financially supporting our book project. Fourth, we thank Kimberley Morris, Harry Fabian and Finn Halligan for their editorial support. Finally, our thanks go to the (former) chairs of our research groups – Tanya Bondarouk (Human Resource Management), Ariana Need and Ringo Ossewaarde (Public Administration) and Michiel Heldeweg (Law, Governance and Technology) for their trust and the possibility to work on our joint book project.

Jeroen Meijerink, Giedo Jansen and Victoria Daskalova
University of Twente, the Netherlands

PART I

Setting the stage – platform-mediated gig work in context

1. Platform economy puzzles: the need for a multidisciplinary perspective on gig work

Jeroen Meijerink, Giedo Jansen and Victoria Daskalova

WHAT'S IN A NAME? WHY THIS BOOK CENTERS AROUND 'PLATFORM-ENABLED GIG WORK'

Platform-enabled gig work, that is, fixed-term activities that organizations and consumers outsource to independent contractors with the help of online labour platforms (Kuhn and Maleki 2017; Meijerink and Keegan 2019), is on the rise (Friedman 2014; Kässi and Lehdonvirta 2018; Kenney and Zysman 2016; Pesole et al. 2018). In the United Kingdom alone, more than 1.1 million individuals are reported to find gig-based work via online labour platforms, such as Uber, Deliveroo and Amazon's Mechanical Turk (Gaskell 2018) – a number which is expected to rise exponentially in the future (ING-Bank 2018). Although providing benefits, such as new business development and business models, flexibility and empowerment for marginalized workers, gig work mediated by online platforms also leads to reduced (social) security for workers, precarious working conditions (including tight algorithmic control and immediate deactivation), digital divides and limited possibilities for collective action (Daskalova 2018; Frenken et al. 2020; Gandini 2019; Meijerink et al. 2019; Rosenblat 2018; Veen et al. 2020; Wood et al. 2019), which make it the subject of intense debates in policy circles and, more broadly, in society. It is precisely these puzzles, related to the opportunities and challenges associated with platform-enabled gig work, which are central to this edited volume.

Delving into these puzzles requires a clear demarcation of the phenomenon of interest to this book. The notion of platform-enabled gig work is inherently puzzling itself. This definitional puzzle becomes apparent in the variety of terms that have been coined to describe recent developments in labour markets and the rise of online labour platforms. Among others, these developments have been referred to as the gig economy (Friedman 2014; Meijerink and

Keegan 2019; Wood et al. 2019), the platform economy (Kenney and Zysman 2016), the on-demand economy (Berg 2015; Shapiro 2018), the Uber economy (Daskalova 2018), work on demand via app (Aloisi 2016), gig work (Stanford 2017), app work (Duggan et al. 2020), crowdwork (De Stefano 2015), platform labour (Van Doorn 2017), digital labour (Fish and Srinivasan 2012; Scholz 2012), eLancing (Aguinis and Lawal 2013) and micro entrepreneurship (Kuhn and Maleki 2017).

For this book, we decided to focus on Meijerink and Keegan's (2019) term 'platform-enabled gig work'. This is not an attempt to synthesize all developments associated with the rise of online labour platforms into one overarching concept. We deliberately chose to focus on platform-enabled gig work since this term ties together the ideas that underlie the different perspectives on online labour platforms, their workers and type of economic exchanges they bring about. Specifically, the term describes the activities that manifest themselves at the intersection of gig work and platform work (or the gig economy and platform economy as economic systems in which work-related activities take place) (see Figure 1.1). For the purpose of this book, we define platform-enabled gig work as fixed-term, paid activities that are executed by independent contractors, outsourced by organizations and/or consumers, and intermediated by online labour platforms. This definition has several implications, as discussed next.

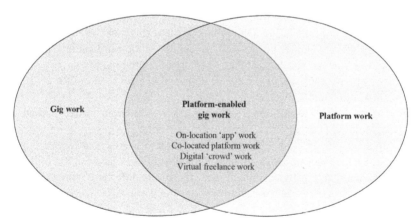

Figure 1.1   Conceptualizing platform-enabled gig work

Independent Work Performed by Freelance Gig Workers

Platform-enabled gig workers are in theory independent and, therefore, perform work outside the confines of the standard employment relationship

(Daskalova 2018; De Stefano 2015; Duggan et al. 2020; Meijerink and Keegan 2019). That is, gig workers are not employed by an organization, but instead perform paid work as freelancers, independent contractors or solo self-employed workers (Aloisi 2016; Jansen 2017; Kuhn and Maleki 2017; Rosenblat 2018). As we see in later chapters, the legal status of many gig workers is disputed and drives many of the public and policy debates around this new type of work. From a historical perspective, this type of freelance working is far from new. For instance, as noted by Stanford (2017), contemporary forms of gig work strongly resemble the on-demand and casual work (for example, as performed by dock workers) that characterized labour markets in the nineteenth century. Moreover, freelancers now are paid per gig performed, resembling the piece-work pay that characterized 'a merchant distributed production tasks to paid employees, supplying necessary raw materials and supplies' that paid workers had to process into (semi-) finished products, such as 'textile, clothing, footwear, cutlery, small furnishings and other simple consumer good' (Stanford 2017, p. 385).

Intermediation by an Online Platform

Freelance activities that are enabled by online labour platforms are the gig work central to this edited volume. Here, online labour platforms refer to organizations that rely on the Internet and related technologies to match-make between supply and demand for labour (Kuhn and Maleki 2017; Meijerink and Keegan 2019; Stanford 2017). Online labour platforms make use of platform technology (for example, software algorithms, online applications (apps), artificial intelligence and big databases) to create online marketplaces where freelance gig workers and those who request freelance services are able to interact (Aguinis and Lawal 2013; Duggan et al. 2020; Rosenblat 2018). This makes online labour platforms a labour market intermediary that mediates between solo self-employed workers who are willing to perform activities, and organizations (in the example of business-to-business matchmaking) and/or consumers (business-to-consumer matchmaking) that wish to contract these services. Nevertheless, online labour platforms differ from other labour market intermediaries, such as temp agencies and social network sites (for example, LinkedIn). For instance, as noted by Meijerink and Keegan (2019), while temporary jobs (temp) agencies establish an employment relationship with the workers they supply to client organizations, online labour platforms do not since they work with independent contractors. Instead, online labour platforms, at least in theory, act as brokers between freelancers and those who request freelance services. Although labour market intermediaries, such as social network sites, headhunting agencies or online job boards, provide matchmaking services (Bonet et al. 2013), unlike online labour platforms they typically do not exert control over labour processes. The involvement of social network sites, headhunting agencies or online job boards ends when a match is made between a worker and an organization. Online labour platforms' involvement lasts longer and is more pervasive (Gandini 2019; Shapiro 2018; Veen et al. 2020) as they control gig workers by means of work practices, such as training, staffing, compensation, appraisal, incentives and task allocation (Cassady et al. 2018; Duggan et al. 2020; Meijerink and Keegan 2019).

Online labour platforms exercise control and automate work practices by means of platform technology. This involves the use of software algorithms for dispatching orders, automated deactivation, online review schemes, surveillance and/or surge pricing (Duggan et al. 2020; Kellogg et al. 2020; Rosenblat 2018; Stanford 2017; Veen et al. 2020). In so doing, algorithms augment, and at times replace, human managers in the execution of managerial activities (Raisch and Krakowski 2021). We realize that employees in employment relationships are also increasingly controlled by the same type of technologies used for controlling the labour processes of freelance platform workers (Gal et al. 2020; Leicht-Deobald et al. 2019; Strohmeier and Piazza 2015). Indeed, although employed, the meal deliverers of Takeaway.com and Foodora nevertheless perform platform work that is controlled by an algorithmic manager (Newlands 2020). Platform work is not unique to online labour platforms that match-make between freelancers and organizations/consumers. Therefore, in order to establish the research scope of the book, this edited volume centres around platform-enabled gig work as a special type of platform work performed by independent contractors. We also realize that freelancers can acquire assignments through means other than online labour platforms (for example, through their personal network). This self-acquired type of gig work is also beyond the scope of this book. Therefore, we focus on platform-enabled gig work only as freelance activities that are intermediated by online labour platforms.

Fixed-term Activities: A Classification of On-demand Work via Online Labour Platforms

In line with the freelance status of gig workers, platform-enabled gig work amounts to activities that are performed on demand (Aloisi 2016; Shapiro 2018; Van Doorn 2017). These activities can be manifold and include, among others, the delivery of food (for example, Deliveroo, Uber Eats and Instacart), driving taxi rides (Uber Rides and Lyft), cleaning (Helpling), image tagging (Amazon Mechanical Turk), programming (Upwork and Fiverr), tour guiding (Hi Hi Guide), bartending (Temper), health care (DoctoronDemand.com), babysitting (Sittercity) or consulting (GigNow and Twago). In line with the work of others (De Stefano 2015; Duggan et al. 2020; Nakatsu et al. 2014), we conceptualize platform-enabled gig work along two dimensions (see Figure 1.2). First, platform-enabled activities differ in the degree to which these are structured (Nakatsu et al. 2014). Structured tasks are those for which a solution to a problem or desired end-state is clearly defined. These tasks typically involve unskilled labour, job simplification, job specialization and limited worker autonomy (Wood et al. 2019). These structured activities are strictly controlled by the online labour platform and broken down into small core elements, such as the delivery of meals (Griesbach et al. 2019; Irani and Silberman 2013), image tagging (Irani and Silberman 2013) or driving taxi rides (Rosenblat 2018). Unstructured tasks are characterized by having no (clearly defined) outcome (Nakatsu et al., 2014). For executing these forms of labour, platform-enabled gig workers are provided with significant levels of job autonomy that offer the freedom and flexibility to find novel solutions (Kuhn and Maleki 2017). Platform-enabled gig work that is relatively unstructured includes graphic design, consultancy and health-care services.

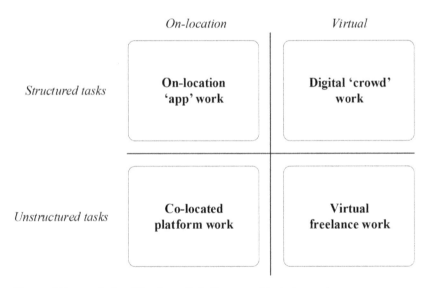

Figure 1.2   A classification of platform-enabled gig work

Second, platform-enabled gig work can be performed on location as well as virtually (Aloisi 2016; Duggan et al. 2020). In the former, the work takes place in the 'real' world. Accordingly, on-location platform work always takes place in physical interaction with others and may, depending on the complexity of the task involved, require gig workers to collaborate with other gig workers and/or employees of the hiring organization (Nakatsu et al. 2014). On-location activities can be both structured – what we term on-location 'app' work such as meal deliveries (Deliveroo) and driving taxi rides (Uber) – as well as unstructured – which we refer to as co-located platform work, such as consultancy services at a client's premises (for example, Twago and GigNow).

Virtual work is performed remotely from the client's premises (Jarrahi and Sutherland 2019; Lehdonvirta et al. 2019; Wood et al., 2019). Indeed, virtual gig workers and their clients are often located in different countries (Pesole et al. 2018) and rely on information technology to collaborate online (Kinder et al. 2019). Although working remotely, virtual gig workers do not necessarily work in isolation, particularly when they and their clients are mutually interdependent (Nakatsu et al. 2014). The latter occurs in virtual freelance work (see Figure 1.2) where gig workers work remotely yet have to collaborate with their clients to perform unstructured tasks (for example, graphic design via the Upwork or Fiverr platforms). Virtual work that involves the execution of structured tasks is referred to as digital crowd work, such as image tagging or online translation services (for example, via Amazon Mechanical Turk).

Platform-enabled Gig Work: Beyond Mere Matchmaking

Platform-enabled gig work involves at least three actors – workers (providers), clients (requesters) and online labour platforms (intermediaries) – who offer services to another (Meijerink et al. 2019) and are situated in wider social, economic, political and legal contexts (Frenken et al. 2020). Similar to traditional business associations (Brandl and Lehr 2019), online labour platforms (proactively) shape the social, political and legal contexts in which they operate, through lobbying activities, mobilizing their platform users for shaping public opinion (Van Doorn 2020) and/or utilizing public policies to their advantage (Meijerink et al. 2019; Uzunca et al. 2018). Moreover, online labour platforms create and/or reconfigure different markets, including: product markets (for example, food delivery, transportation, hospitality, cleaning and consultancy), labour markets where online labour platforms formalize, organize and match-make between supply and demand for labour (for example, organizing dispersed and informal labour markets), markets for intermediation services where online labour platforms compete for market share (for example, Uber, Lyft and Juno attempting to become monopolist in their respective market for intermediation), information market (that is, selling data or offering consultancy services) and, at times, financial markets (for example, taxi platforms such as Grab and Gojek which start offering financial services to their users). These markets have implications for workers, consumers, governments, organizations, labour unions, incumbent labour market intermediaries, retailers, service organizations and citizens generally. Indeed, sitting at the intersection of platform work and gig work (see Figure 1.1), platform-enabled gig work creates new challenges and reinforces existing challenges for societal stake-holders in unprecedented ways (for example, controlling freelancers, reclassi-fication issues, precarity, power imbalances and privacy). It is these challenges that this edited volume addresses.

OVERALL AIM OF THE BOOK: WHY A MULTIDISCIPLINARY APPROACH IS NEEDED

The idea of a self-clearing marketplace for labour where supply meets demand is brilliant in its simplicity. Persons with resources (time, skills or spare assets) are matched with other persons (or companies) in need of someone with a par-ticular profile. The market clears. Consumers benefit from increased choice, better matches, higher quality and, possibly, lower prices thanks to greater competition; workers enjoy more opportunities in relation to quantity, choice of different customers and potential for higher earnings. Thanks to speedy Internet connections and clever algorithms, the platform helps solve problems of coordination and allocates tasks efficiently.

Yet online labour platforms are not just the latest brilliant solution to what is fundamentally an economic problem and, arguably, a problem only for specific sectors. The online labour platform phenomenon is a game changer for labour markets generally; that is, the type of innovation which promises to disrupt (or is expected to disrupt) established ways of organizing and manag-ing workplaces, hiring and firing, consuming and producing, and interacting and organizing. This digital revolution, possibly one of many to come, is set to upend established institutional patterns in management, social dialogue and industrial relations, labour law and social policy. This trend, and the responses it generates, raises not only practical, implementation-related questions (for example, how to ensure that self-employed workers contribute to a pension fund), but also triggers normative concerns. That is, debates concerning, for instance, Uber and its competitors are not only about making the taxi industry more efficient; they are more generally about what labour and labour institu-tions ought to be like in society.

Online labour platforms create or complicate a number of difficult problems for policy-makers, companies, workers and society as a whole. The nature of these problems, referred to as 'puzzles' in this book, have provided the inspiration for this multidisciplinary project. Delving into some of them briefly helps explain the editorial choices we have made as regards the scientific perspectives included, the selection of authors and topics, and the structure of the chapters. They reflect our concern with framing the gig economy develop-ments in pragmatic, problem-driven terms, disconnected from the theoretical advances and historical experience which has accumulated over at least two centuries.

Platform Economy Puzzles

One puzzle is the insider–outsider problem, a growing problem for labour institutions (Jansen and Lehr 2019; Lindbeck and Snower 1984; Rueda 2007). This phenomenon precedes the development of the digitally mediated gig economy; arguably, it is at the crux of the success (and fears) related to the gig economy. The insider–outsider debate fundamentally concerns the conflict between the interest of those with secure and protected jobs (insiders) and those without (outsiders), and how institutional choices create perverse incentives which exacerbate this tension (Rueda 2007). Labour institutions (including trade unions) are not able to serve all persons, nor regulate all services provided by those persons – typically, labour institutions define their target group as those who find themselves in an employment relationship. By definition this is a narrower group than the total number of people performing work in one way or another. Insiders, that is, those with a stable and well-paid job, and their representatives, arguably have limited incentives to improve opportunities for outsiders, that is, those persons looking for a job (Lindbeck and Snower 1984). By protecting only those who are already in possession of employment, the argument goes, labour institutions limit opportunities for others who wish to enter the labour market (and especially those in need of flexibility) (Emmenegger 2014). The independence, empowerment and direct access to income opportunities promised by platform companies, speaks directly to the needs of marginalized workers and especially those who are (systematically) excluded by the insider-dominated traditional labour markets.

Labour platforms solve this problem by lowering the barriers to entry of labour markets (or, some would argue, the barriers to entry for entrepreneurship; Daskalova 2018). They expand opportunities for participation in (online) labour markets. By circumventing the standard employment relationship dominated by insiders, they open up opportunities for outsiders. Partially, this is thanks to the flexibility offered by technology, which enables work to be combined with other activities or time limitations: studying, caring for children or sick relatives, or undertaking extensive medical treatment. However, the image of the platform as a democratizing force of opportunities needs to be critically examined. Does platform work solve the insider–outsider problem, or make it worse, by normalizing a second-class workforce which works under more precarious conditions for less income than insiders earn?

The insider–outsider debate leads us to another puzzle of labour market institutions, which may be termed the representation paradox. Media reports suggest that platform workers have fewer rights, greater responsibilities and lower income than their regularly employed counterparts. Despite this, platform workers shy away from union membership, a phenomenon which seems hard to square with the role that unions are expected to play on labour markets.

In theory, unions pursue solidarity among workers (all workers, regardless of status) and are tasked to take action to improve the labour conditions of all workers, not only those among their membership. Paradoxically, unions, whose membership has been declining in the past decade, have shown some reluctance to represent atypical workers (Jansen and Lehr 2019), which often include platform workers. In response to this, alternative, grass-root forms of worker representation and organizing are emerging – both within the platforms themselves and in the form of worker cooperatives (Scholz 2012). Yet how likely are these alternative forms of representation to succeed? Can we expect to see new actors emerge, or even replace, established players in the field of industrial relations? A look back to history and theory, combined with insights from data on the gig economy, can help. Online labour platforms expose the difficulties inherent in organizing workers whose interests differ and may even be diametrically opposed. After all, labour markets are populated by individuals with different skills, needs and limitations (including personal obligations and geographic locations) who nonetheless all compete for generally standardized occupations. These difficulties in organizing suggest a fragmentation of representative institutions, while the very presence of labour platforms undermine the legitimacy and thereby weaken the power of regular unions. This raises the question of what the future of social dialogue and industrial relations will look like, and whether the creative destruction ushered in by the platform revolution will replace old actors with new actors, or simply upend the existing arrangements without generating viable alternatives.

Turning to the nature of platform work itself and how it is experienced by individual workers, there are also puzzling inconsistencies between narrative and reality to be found. Technology-optimistic narratives emphasize that they bring freedom and empowerment to platform workers owing to flexibility in scheduling, the absence of a manager ('Be your own boss!') and increased autonomy ('You decide when and how much to work'). However, media reports and academic publications have quickly dispelled these claims (Aloisi 2016; Rosenblat 2018; Wood et al. 2019). A physically present human boss may be oppressive but, apparently, so is the ubiquitous presence of an invisible algorithm. Whereas the traditional systems of control linking worker to management are virtually erased, new and less well known forms of control – algorithmic management and user reviews – are now in control (Kellogg et al. 2020). The management of work may thus only seem free from human bias in hiring, firing and workplace management, but the reality is that the algorithm is not flawless; multiple studies note evidence of algorithms reproducing societal hierarchies and biases (Faraj et al. 2018; Leicht-Deobald et al. 2019; Pasquale 2015; Zuboff 2019). Being your own boss is attractive as regards making decisions, but it also comes with greater responsibility for your own financial and legal liabilities (such as responsibility for worker accidents or damage to a client's property). Workers may seem independent; however, they are embedded in a particular technological and business ecosystem (Frenken et al. 2020; Friedman 2014). The flattening of this organization is anything but apparent. For workers attracted to the idea of entrepreneurship and autonomy, this realization raises the question of empowerment: to what extent does their freelance status help them advance, or does it also emasculate them? One of the paradoxes of labour platforms lies in this guarded promise toward workers to enhance their autonomy by subjecting them to algorithmic management (Gandini 2019; Shapiro 2018). It is worth asking, is the algorithm gentler, more generous and more forgiving than a flesh-and-blood manager? Similarly, the supposed freedom brought about by platforms may be questioned, given the need for income, and the conditional and uncertain nature of income. The technology needs to be seen in context. As an old joke goes, 'The perks of the job: you can work whenever you want; every day is a Saturday; the cons: you work on Saturday'.

The Objective of this Volume

The changes outlined previously are important; all the more so because they are far reaching. Labour market institutions and concepts such as labour law, collective bargaining and the notion of human resource management are under pressure owing to these developments in online labour platforms (Aloisi 2016; Berg 2015; De Stefano 2015; Frenken et al. 2020; Meijerink and Keegan 2019). Perceptions and experiences of individuals (regarding empowerment, precariousness and representation) shape responses from workers and employers, but also from policy-makers and citizens. Currently, research into platform-enabled gig work is dispersed across a number of fields, such as law, labour economics, sociology, history and technology, geography, new media studies, business, human resource management, work and organizational psychology, and entrepreneurship (Sutherland and Jarrahi 2018). However, the puzzles associated with platform-enabled gig work are far from mono-disciplinary in nature. The same applies to reaping the benefits of gig work and turning its potential into real value for workers, platform firms and other societal stakeholders. Policy responses as well as academic literature on the topic often tend to focus on national circumstances, and often on specific issues which need to be addressed with some urgency. However, the changes in labour markets are deeper and call for a broader perspective, a sense of context and a comparative perspective. Explaining the gig work phenomenon and designing responses to it requires knowledge of these different labour market institutions and their interaction with platforms and platform workers. This requires a multidisciplinary perspective on the drivers and nature of the puzzles associated with platform-enabled gig work, and the implications (both desired and undesired) for societies, labour markets, public policy organizations and individuals. Accordingly, this book embarks on a perilous task by offering a multidisciplinary perspective on platform-enabled gig work and the implications (both intended and unintended) this has for societies, labour markets, public policy organizations and individuals. We are certain to miss some important disciplinary perspectives and we can only offer snippets from different disciplinary contributions, but we nonetheless believe we have taken an important step forward in opening the space for multidisciplinary dialogue on the topic of platform work.

APPROACH AND OUTLINE OF THE BOOK

This volume takes stock of and synthesize insights from different academic disciplines on past and future challenges as well as opportunities associated with platform-enabled gig work. In doing so, the book explores whether established theories are useful to understand the major puzzles of the platform economy, and how theory, research and policy on platform-enabled gig work may be advanced. Bringing together different strands of research will provide a bird's-eye view of developments and will enable fresh perspectives on the possibilities and problems faced by policy-makers, unions, industry and business, and, more broadly, society.

To ensure complementarities across the different chapters, and to foster readability, each chapter is structured similarly, whereby each chapter reflects on the following issues as seen from the selected academic disciplines that study gig work. First, each chapter revolves around a specific problem (a puzzle) that reflects either an urgent societal issue or an academic question. Second, each chapter discusses the causes and consequences of the problem at hand. Based on these insights, the authors discuss potential solutions to either fill the knowledge gap in the literature and/or offer practical policy recommendations. Finally, each chapter presents directions for future research.

This volume consists of three parts. Part I sets the stage by placing platform-mediated gig work in a broader context. First, in Chapter 2, Annarosa Pesole discusses the conceptual and methodological problem of measuring platform-enabled gig work in labour economics. Empirically investigating platform labour is, in the words of Huws et al. (2017, p. 52) 'a step into the unknown' as there is no consensus on definitions and no existing baseline data nor any established method of collecting it. The basic question of how prevalent platform-mediated gig work really is, therefore is more complex than it seems. To date, most data on platform work is anecdotal or relies on small-scale qualitative studies. In particular, quantitative and comparative data on platform labour are limited. Pesole's chapter discusses the advantages and disadvantages of different approaches for collecting data on the prevalence and nature of platform work. After discussing measurement approaches, she reports findings on the size and nature of platform work in Europe based on data from the COLLEEM survey by the European Commission's Joint Research Centre (JRC).

In Chapter 3, Jim Stanford raises the issue of how novel platform-mediated gig work really is. Despite being promoted as a technologically-driven and innovative work practice, Stanford provides a historical perspective on platform labour and argues that the core features of digitally mediated gig work are hundreds of years old, having been applied in pre-digital modes of capitalism. In this chapter, Stanford argues that gig work is better understood as a return to previous practices, instead of a new model of employment, and is part of a wider increase in insecure and non-standard work under neoliberalism. He also argues that not just technology, but also economic, political and social determinants condition the prevalence of insecure and non-standard labour.

In Chapter 4, Victoria Daskalova, Shae McCrystal and Masako Wakui take a legal perspective. They argue that platform-mediated gig work revives some old problems regarding regulating labour markets while simultaneously challenging existing legal solutions to those problems. One of the key puzzles in the platform economy is to do with legal classifications: are platform workers independent contractors or are they employees in disguise? This chapters discusses the tension between two legal fields. On the one hand, is labour law and collective bargaining, the traditional legal solutions to protect workers against social and labour risks. On the other, is commercial law, including competition law. The latter, although originally developed to fight concentrations of power in cartels, often raises barriers to collective bargaining by self-employed workers.

Part II of this book addresses puzzles in economic and social exchanges in the platform-enabled gig economy. In Chapter 5, Aaron Shapiro addresses the question of why urban areas are an attractive source of (economic) value for online platform firms. He discusses how platforms utilize the accessibility and collective utility of urban infrastructures. Building on urban theory, Shapiro introduces the notion of infrastructural surplus. In particular, he discusses two modes by which platforms may extract value from urban infrastructures: reformatting social space and transactional exclusion. The former involves platforms directly reformatting social space as an infrastructural support for their operations; the latter captures infrastructural surplus indirectly, by excluding resources necessary to perform gigs from the platform-mediated labour transaction.

Chapter 6 by Niels van Doorn and Adam Badger also focuses on value extraction. The puzzle they address is how gig economy companies can continue to grow despite regularly suffering substantial losses. To conceptualize this puzzle, Van Doorn and Badger introduce the notion of dual value production, which describes how platforms capture two types of value from gig work: the monetary value associated with the service transaction, and the more speculative value associated with the data generated during service provision by gig workers. By elaborating on the construction of data as a specific asset class, this chapter examines platform-enabled gig work from the perspective of political economy of data and finance capital.

In Chapter 7, Anne Keegan and Jeroen Meijerink discuss how gig work mainly takes place outside the confines of an employment relationship, while most gig workers are nevertheless subject to a range of human resource management (HRM) activities (such as recruitment, selection, appraisal, compensation and job design) that traditionally are seen to uphold employment relationships. It is this puzzle – HRM activities without employment relationships – that is central to this chapter. From the perspective of HRM scholarship, Keegan and Meijerink examine why the business model of online labour platforms, and the platform ecosystems they create, requires the use of HRM activities and how this creates institutional complexity. Moreover, they explore the consequences of the institutional complexity for gig workers, platforms and societal stakeholders.

Chapter 8 by James Duggan, Ultan Sherman, Ronan Carbery and Anthony McDonnell focuses on the multi-party working relationships between platform firms, gig workers and consumers or hiring organizations. The problem this chapter starts out with is that in platform labour the traditional concept of a legal employment relationship between an employer and employee is increasingly less applicable. Drawing from the fields of work and organizational psychology, Duggan and colleagues adopt an alternative perspective to understand employment relationship in the platform economy: the psychological contract. Psychological contract theory examines the mutual promise-based expectations that parties have of one another and how these implicit expectations impact behaviour. Using psychological contract theory, this chapters examines the (relational and transactional) expectations that may exist in the working relationships between gig workers, the platform firms for whom they work, and consumers or hiring organizations.

Part III of this book explores the wider social and political implications of platform-mediated gig work. In particular, this part of the book looks into collective organization and interest representation. First, in Chapter 9, Damion Jonathan Bunders examines the puzzle of collective action. This chapter builds on insights from sociology and social history to examine how gig workers themselves might strive for better work. Yet, while standard sociological theory would dictate that individualized work practices of platform workers hinder collective organization, Bunders shows that various forms of collective action by gig workers can be observed in practice. As an extreme case of collective action, this chapter discusses worker-owned platform cooperatives.

Bunders compares worker-owned gig platforms with traditional worker cooperatives in order to derive insights on the chances for collective action in the platform economy.

Chapter 10, shifts the focus from self-organization to interest representation. In this chapter Paul Jonker-Hoffrén and Giedo Jansen discuss how digital platform-based work challenges existing models of interest representation, which are often based the notions of place-based citizenship (that is, political representation) and/or type-of-contract segmentation (that is, representation in a country's systems of industrial relations). Combining insights from political science and industrial relations, Jonker-Hoffrén and Jansen argue that representation of platform workers is a function of both demand for and supply of representation. They present a framework, based on insider–outsider theory, through which representation by political parties, trade unions and other interested organizations can be studied.

Finally, in Chapter 11, we as editors synthesize the book by integrating insights on the causes, consequences and (policy) solutions to the various platform economy puzzles as outlined in the different chapters of this volume. We also identify avenues for future interdisciplinary research into the challenges and opportunities of platform-enabled gig work. To ensure the implementation of our proposed research avenues, we conclude with a discussion of methodologies and data collection techniques for new research into platform economy puzzles.

REFERENCES

Aguinis, H. and S.O. Lawal (2013), 'eLancing: a review and research agenda for bridging the science–practice gap', *Human Resource Management Review*, **23** (1), 6–17.
Aloisi, A. (2016), 'Commoditized workers: Case study research on labor law issues arising from a set of on-demand/gig economy platforms', *Comparative Labor Law and Policy Journal*, **37** (3), 653–90.
Berg, J. (2015), 'Income security in the on-demand economy: findings and policy lessons from a survey of crowdworkers', *Comparative Labor Law and Policy Journal*, **37** (3), 543–76.
Bonet, R., P. Cappelli and M. Hamori (2013), 'Labor market intermediaries and the new paradigm for human resources', *Academy of Management Annals*, **7** (1), 341–92.
Brandl, B. and A. Lehr (2019), 'The strange non-death of employer and business associations: an analysis of their representativeness and activities in Western European countries', *Economic and Industrial Democracy*, **40** (4), 932–53.
Cassady, E.A., S.L. Fisher and S. Olsen (2018), 'Using eHRM to manage workers in the platform economy', in J.H. Dulebohn and D.L. Stone (eds), *The Brave New World of eHRM 2.0*, Charlotte, NC: Information Age, pp. 217–46.
Daskalova, V. (2018), 'Regulating the new self-employed in the Uber economy: what role for EU competition law', *German Law Journal*, **19** (3), 461–508.

De Stefano, V. (2015), 'The rise of the just-in-time workforce: on-demand work, crowdwork, and labor protection in the gig-economy', *Comparative Labor Law and Policy Journal*, **37** (3), 471–504.

Duggan, J., U. Sherman, R. Carbery and A. McDonnell (2020), 'Algorithmic management and app-work in the gig economy: a research agenda for employment relations and HRM', *Human Resource Management Journal*, **30** (1), 114–32.

Emmenegger, P. (2014), *The Power to Dismiss: Trade Unions and the Regulation of Job Security in Western Europe*, New York: Oxford University Press.

Faraj, S., S. Pachidi and K. Sayegh (2018), 'Working and organizing in the age of the learning algorithm', *Information and Organization*, **28** (1), 62–70.

Fish, A. and R. Srinivasan (2012), 'Digital labor is the new killer app', *New Media and Society*, **14** (1), 137–52.

Frenken, K., T. Vaskelainen, L. Fünfschilling and L. Piscicelli (2020), 'An institutional logics perspective on the gig economy', in I. Maurer, J. Mair and A. Oberg (eds), *Theorizing the Sharing Economy: Variety and Trajectories of New Forms of Organizing*, Bingley: Emerald.

Friedman, G. (2014), 'Workers without employers: shadow corporations and the rise of the gig economy', *Review of Keynesian Economics*, **2** (2), 171–88.

Gal, U., T.B. Jensen and M.-K. Stein (2020), 'Breaking the vicious cycle of algorithmic management: a virtue ethics approach to people analytics', *Information and Organization*, **30** (2), art. 100301.

Gandini, A. (2019), 'Labour process theory and the gig economy', *Human Relations*, **72** (6), 1039–56.

Gaskell, A. (2018), 'The Demographics of the gig economy', *Forbes*, 1 August, accessed 9 December 2020 at https://www.forbes.com/sites/adigaskell/2018/08/01/the-demographics-of-the-gig-economy/#6d7306b969fb.

Griesbach, K., A. Reich, L. Elliott-Negri and R. Milkman (2019), 'Algorithmic control in platform food delivery work', *Socius*, **5** (1), 1–15.

Huws, U., N. Spencer, D.S. Syrdal and K. Holts (2017), *Work in the European Gig Economy: Research Results from the UK, Sweden, Germany, Austria, the Netherlands, Switzerland and Italy*, Brussels: Foundation for European Progressives Studies.

ING-Bank (2018), 'Platformen Kunnen Arbeidsmarkt Drastisch Veranderen' ('Platforms can change the labor market drastically') accessed 9 December 2020 at https://www.ing.nl/media/pdf-EBZ-Platformen%20kunnen%20arbeidsmarkt%20drastisch%20veranderen_tcm162-159443.pdf.

Irani, L.C. and M.S. Silberman (2013), 'Turkopticon: interrupting worker invisibility in Amazon Mechanical Turk', in *Proceedings of the SIGCHI Conference on Human Factors in Computing Systems (CHI '13)*, New York: Association for Computing Machinery, pp. 611–20, doi:10.1145/2470654.2470742.

Jansen, G. (2017), 'Farewell to the rightist self-employed? "New self-employment" and political alignments', *Acta Politica*, **52** (3), 306–38.

Jansen, G. and A. Lehr (2019), 'On the outside looking in? A micro-level analysis of insiders' and outsiders' trade union membership', *Economic and Industrial Democracy*, 23 December, accessed 12 April 2021 at https://doi.org/10.1177%2F0143831X19890130.

Jarrahi, M.H. and W. Sutherland (2019), 'Algorithmic management and algorithmic competencies: understanding and appropriating algorithms in gig work', in N. Taylor, C. Christian-Lamb, M. Martin and B. Nardi (eds), *Information in Contemporary Society*, Cham: Springer, pp. 578–89.

Kässi, O. and V. Lehdonvirta (2018), 'Online labour index: measuring the online gig economy for policy and research', *Technological Forecasting and Social Change*, **137** (December), 241–8.

Kellogg, K.C., M.A. Valentine and A. Christin (2020), 'Algorithms at work: the new contested terrain of control', *Academy of Management Annals*, **14** (1), 366–410.

Kenney, M. and J. Zysman (2016), 'The rise of the platform economy', *Issues in Science and Technology*, **32** (3), 61–9.

Kinder, E., M.H. Jarrahi and W. Sutherland (2019), 'Gig platforms, tensions, alliances and ecosystems: an actor-network perspective', *Proceedings of the ACM on Human-Computer Interaction*, **3** (November), art. 212, doi:10.1145/3359314.

Kuhn, K.M. and A. Maleki (2017), 'Micro-entrepreneurs, dependent contractors, and instaserfs: understanding online labor platform workforces', *Academy of Management Perspectives*, **31** (3), 183–200.

Lehdonvirta, V., O. Kässi, I. Hjorth, H. Barnard and M. Graham (2019), 'The global platform economy: a new offshoring institution enabling emerging-economy microproviders', *Journal of Management*, **45** (2), 567–99.

Leicht-Deobald, U., T.Busch, C. Schank, A. Weibel, S. Schafheitle, I. Wildhaber, et al. (2019), 'The challenges of algorithm-based HR decision-making for personal integrity', *Journal of Business Ethics*, **160** (2), 377–92.

Lindbeck, A. and D.J. Snower (1984), *Involuntary Employment as an Insider-Outsider Dilemma*, Stockholm: Institute for International Economic Studies.

Meijerink, J.G. and A. Keegan (2019), 'Conceptualizing human resource management in the gig economy: toward a platform ecosystem perspective', *Journal of Managerial Psychology*, **34** (4), 214–32.

Meijerink, J.G., A. Keegan and T. Bondarouk (2019), 'Exploring "human resource management without employment" in the gig economy: how online labor platforms manage institutional complexity', paper presented at the Sixth International Workshop on the Sharing Economy, Utrecht, 28–29 June.

Nakatsu, R.T., E.B. Grossman and C.L. Iacovou (2014), 'A taxonomy of crowdsourcing based on task complexity', *Journal of Information Science*, **40** (6), 823–34.

Newlands, G. (2020), 'Algorithmic surveillance in the gig economy: the organisation of work through Lefebvrian conceived space', *Organization Studies*, 9 July, accessed 2 April 2021 at https://doi.org/10.1177%2F0170840620937900.

Pasquale, F. (2015), *The Black Box Society*, Cambridge, MA: Harvard University Press.

Pesole, A., M. Brancati, E. Fernández-Macías, F. Biagi and I. González Vázquez (2018), *Platform Workers in Europe*, Luxembourg: Publications Office of the European Union.

Raisch, S. and S. Krakowski (2021), 'Artificial intelligence and management: the automation-augmentation paradox', *Academy of Management Review*, **46** (1), 192–210.

Rosenblat, A. (2018), *Uberland: How Algorithms Are Rewriting the Rules of Work*, Oakland, CA: University of California Press.

Rueda, D. (2007), *Social Democracy Inside Out: Partisanship and Labor Market Policy in Advanced Industrialized Democracies*, Oxford: Oxford University Press.

Scholz, T. (2012), *Digital Labor: The Internet as Playground and Factory*, New York: Routledge.

Shapiro, A. (2018), 'Between autonomy and control: strategies of arbitrage in the "on-demand" economy', *New Media and Society*, **20** (8), 2954–71.

Stanford, J. (2017), 'The resurgence of gig work: historical and theoretical perspectives', *Economic and Labour Relations Review*, **28** (3), 382–401.

Strohmeier, S. and F. Piazza (2015), 'Artificial intelligence techniques in human resource management – a conceptual exploration', in J. Kacprzyk and L. Jain (eds), *Intelligent Techniques in Engineering Management*, New York: Springer, pp. 149–72.

Sutherland, W. and M.H. Jarrahi (2018), 'The sharing economy and digital platforms: a review and research agenda', *International Journal of Information Management*, **43** (December), 328–41.

Uzunca, B., J.C. Rigtering and P. Ozcan (2018), 'Sharing and shaping: a cross-country comparison of how sharing economy firms shape their institutional environment to gain legitimacy', *Academy of Management Discoveries*, **4** (3), 248–72.

Van Doorn, N. (2017), 'Platform labor: on the gendered and racialized exploitation of low-income service work in the "on-demand" economy', *Information, Communication & Society*, **20** (6), 898–914.

Van Doorn, N. (2020), 'A new institution on the block: on platform urbanism and Airbnb citizenship', *New Media and Society*, **22** (10), 1808–26.

Veen, A., T. Barratt and C. Goods (2020), 'Platform-capital's "app-etite" for control: a labour process analysis of food-delivery work in Australia', *Work, Employment and Society*, **34** (3), 388–406.

Wood, A.J., M. Graham, V. Lehdonvirta and I. Hjorth (2019), 'Good gig, bad gig: autonomy and algorithmic control in the global gig economy', *Work, Employment and Society*, **33** (1), 56–75.

Zuboff, S. (2019), *The Age of Surveillance Capitalism: The Fight for a Human Future at the New Frontier of Power*, New York: Public Affairs.

2. Understanding the prevalence and nature of platform work: the measurement case in the COLLEEM survey study

Annarosa Pesole

INTRODUCTION

Despite the increasing interest on platform work from policy-makers, researchers and the public, still very little is known about the phenomenon of platform work. Currently, reliable estimates on the size of platform work are still missing, and the various attempts carried out by national statistics offices as well as private and academic researchers have highlighted the difficulties of measuring its size and the lack of consensus on a common definition.

The platform economy has exploded in the past 10 years, changing both goods and services markets. The phenomenon of digital labour platforms is slightly younger but has already shown large potential for disrupting the conventional concepts of employment and jobs. Indeed, digital labour platforms have changed entirely the way jobs are searched for, obtained, performed and eventually organised. The effects of this restructuring of employment relationships are not entirely clear (Kenney and Zysman 2016; Kuhn and Maleki 2017; Meijerink and Keegan 2019). On the one hand, many advocate that digital labour platforms, owing to the increased flexibility in the working conditions they offer, may lower entry barriers facilitating the participation in the labour market of previously penalised or excluded categories, such as workers with strong family commitments, people with disabilities or health conditions, those who are not in education, employment or training (NEET), the long-term unemployed and people with a migrant background. On the other hand, the very same increased flexibility of working relationships that platforms entail may increase the risk of unfair treatment, disrupted working careers and lack of social protection and security in the workplace. In order to understand the impact of digital labour platforms on the labour market and society in general, it is necessary to understand their characteristics and the way they work (Pesole et al. 2018).

In this chapter, digital labour platforms are defined as 'digital networks that coordinate labour service transactions in an algorithmic way' (Pesole et al. 2018, p. 7). They allow clients to match the need for a specific task with a worker who possesses the capability to perform that task. The matching takes place online through the intervention of algorithms that pair services to workers, with initial pre-established working conditions. The rules that govern this matching are at the core of a new frontier of work organisation and management: algorithmic management. In most instances, these rules are obscure to both the clients and the workers, who are mostly left with the 'take it or leave it' option offered by the match.

However, algorithmic management goes well beyond offering mere match-making services. It allows companies to supervise a multitude of distributed workers in an optimised manner and on a large scale (Duggan et al. 2020). Algorithms define the rules of work assignation, work process optimisation (for example, the surging price system adopted by Uber) and workers' performance evaluation (for example, the rating system). That is, digital labour platforms may rely on constant algorithmic monitoring to ensure tight control over every aspect of work and service delivery. Acceptance rates, time to perform the task, ratings and other features are factored into the algorithm equation. Deviation or failure to comply with the platform policy can bring about sanctions, even up to a one-sided decision to deactivate the worker's profile.

The reorganisation of work processes that algorithmic management entails is de facto a radical deepening of division of labour. Jobs in the offline labour markets are usually constituted by a set of cross-complementary tasks to which a worker is assigned. Digital labour platform enable a process of jobs specialisation that reduces the labour provision to the accomplishment of a single task. According to Cook et al. (2018, p. 1). digital labour platforms 'divide work into small pieces and then offer those pieces of work to independent workers in real-time with low barriers to entry'. Pesole et al. (2018, p. 8) refer to the phenomenon as 'unbundling of tasks'. This main feature of digital labour platforms is the origin of the term 'gig economy', since the labour provision resembles a unique self-contained performance with no attribute of continuity (Meijerink and Keegan 2019; Stanford 2017). Similarly, unclear working conditions and labour relationships have in other instances led to the misleading concepts of 'sharing' and a collaborative economy (Frenken and Schor 2017).

These definitions are not devoid of relevance. According to the label chosen, different characteristics have been attributed to platforms, with subtle implications for the way they are perceived, measured and, eventually, regulated. The purpose of this chapter is to discuss platform work, analysing its size and characteristics, and providing some basic definitions. I avoid using normatively biased terms and, instead, use relatively neutral terms, such as 'digital labour platforms', 'platform work' and 'platform workers'. Equally, I refer to the term 'task' to identify the gig provided through digital labour platforms.

In the next section, before getting to the core issue of measurement, I briefly present the key policy issues raised by platform work. Then I continue analysing previous attempts at measurement and their associated challenges, and I present the results of the Collaborative Economy and Employment (hereafter COLLEEM) survey. The final section concludes and offers directions for future research.

CHALLENGES FOR POLICIES AND THE NEED FOR MEASURING PLATFORM WORK

Measuring labour force participation is one of the key elements of economic policy. Statistics on employment rates are needed to address public concerns, such as social inclusion, social security sustainability, participation in public life and minimum guaranteed living conditions.

A feature of work in digital labour platforms is its indistinctness, which is particularly problematic for policy purposes. Indeed, the unstructured and atomised nature of platform work, together with the novelty of the phenomenon, makes it very difficult to measure this type of work adequately and raises several policy concerns.

Sound measurements and metrics are imperative for policy-makers to understand the evolution of platform work, assess the potential impact of different policy options and evaluate the efficiency of the actions undertaken. Measurement approaches for platform work currently present a dual gap: methodological gaps and availability gaps. Methodological gaps refer to the lack of common definitions and metrics, the lack of new indicators and the lack of new ways of collecting data. Availability gaps follow as a direct consequence of the former, and refer to the lack of data and the lack of harmonisation and comparability of the few data available. National statistical offices and international organisations are working on the definition of platform work, and are developing guidelines to measure platform work. Nevertheless, what makes the task so arduous? In the following, I try to highlight the most chameleonic and still unclear aspects of platform work and how they affect measurement.

A potential structure of metrics to assess platform work should answer to the questions of how many and how good jobs in platforms are. Here we encounter the first obstacle. Jobs are not easily identifiable on labour platforms, since many of the labour services exchanged via platforms are accomplishment of single tasks (that is, unbundling of tasks). In order to answer to the question of how many, it is important to understand how platforms have changed the basic unit of analysis (from jobs to tasks) and why this has consequences for measurement purposes.

The task specialisation that the unbundling of tasks involves enables a market-based redistribution of resources in the labour market. This reorganisation of labour according to market principles overlooks both labour market institutions and traditional employment relationships. That is, platform workers provide labour services within an unregulated market and do not fit entirely any of the substantive categories of the International Classification of Status in Employment (ICSE). The lack of a legal framework and statistical standards makes it very difficult to identify the right criteria or questions to capture platform work in official labour statistics. In addition, the atomised nature of labour provision via platforms makes it even more complicated to quantify how many platform workers are there, as it should be possible to quantify how many tasks compose a job.

A correct assessment of the number of people engaged in digital labour platform activities, as well as an estimate of those for which platforms are the major source of income, is necessary for governments to gauge the need for policy interventions. Furthermore, knowing the size of platform work will give an estimate of the number of workers currently active in the labour market not covered by a suitable employment law framework.

The lack of an appropriate framework influences the how good question as well. In order to understand the quality of work obtained via platforms, one of the prominent regulatory issues to be assessed is the definition of the employment status of platform workers (see Chapters 7 and 8 in this volume) and the subsequent labour rights. Traditionally, workers have been classified in two categories – employees and self-employed – where each category comes with established labour rights and levels of social protection. Platform workers do not neatly match either the employees or the self-employed category, broadening the span of work arrangements that fall under the umbrella of non-standard work. The legislative void, as well as the existence of power asymmetries in the market not mediated by labour market regulations, may exacerbate the power struggle in the platform economy, shifting the costs and risks from platforms to workers, and putting the latter at a higher risk of exploitation and precariat. From their side, digital labour platforms dismiss any responsibility as employers, and claim that platform workers are mostly own-account workers or independent contractors as their sole obligation is to achieve the results agreed with the clients, but they have in principle the autonomy and the flexibility to carry out the labour as they see fit. However, the range of labour services supplied via platforms varies greatly and so does the nature of the working arrangements. In many digital labour platforms, aspects such as tight control, monitoring and authority of the platform over the labour provision closely resemble a subordinate employment relationship. An accurate measurement and definition of the characteristics of platform work will help to shed light on the grey area that lies between the traditional paradigms of, on the one hand, subordinate employment and, on the other, autonomous self-employment.

Nevertheless, the definition of the employment status of platform workers, as well as their access to minimum labour rights and social protection, is not the unique indicator of quality of platform work. Indeed, platform work raises additional social issues linked to access to training and skills development, tax revenues losses (see Chapter 6 in this volume) and competition law (see Chapter 4 in this volume) for which accurate estimations are indispensable toolkits for governments and policy-makers.

As regards training and skills development, data are needed to monitor how skills demand is developing and diverging in digital labour platforms in respect of traditional labour markets. Similarly, governments need data to predict the risk of deskilling. While in standard employment employers have an incentive to train their workforce in the skills required, the training costs in platform work are entirely shifted to the workers who are also at greater risk of entering a low-productivity trap. Furthermore, data on how skills are matched on the platforms are necessary to inform policy-makers so they may design new education and training policies to address skills gaps, skills mismatch, reskilling, digital skills and other issues in this new context. Although in traditional labour markets publicly regulated qualification systems play an important role in skills matching, in digital labour platforms skills matching relies on platforms' proprietary data and matching algorithms alone, making obsolete the standard tools of skills and education policy (Cedefop 2020). Finally, data are also needed to quantify the effects of what is termed the platform 'lock-in'. That is, many platforms do not allow the portability of the ranking or of the worker's profile, making it more onerous for workers to supply their services in multiple platforms, which indirectly affects competition among digital labour platforms and puts platform workers under the stringent pressure of adapting to the skills demand in each specific platform.

Reliable estimates of platform work are also fundamental in understanding how platform work could affect the tax system. On the one hand, the traceability of each transaction on digital labour platform could favour the reporting of certain activities typically carried out through informal channels. On the other hand, the risk associated with an unclear employment status or bogus self-employment is that employers reduce their own labour costs and the overall tax burden for firms, which in turn results in a loss of public revenues generated.

Finally, considering platform workers as independent contractors, in addition to not to granting several employment rights such as paid sick leave and holydays, maternity and unemployment benefits, impede the right to collective bargaining and freedom of association. Freedom of association is inextricably linked to collective bargaining as well as to the right to strike. The preclusion for platform workers to unionise makes it more difficult for them to respond in times of crisis. At the time of writing, the Covid-19 outbreak is hitting several countries in the world and social distancing has become the norm. According to AppJobs, an online platform that collects and compares data on application (app)-based jobs around the world, the Covid-19 pandemic has led to an increase in the global demand for delivery jobs (AppJobs Institute 2020). Many cities in lockdown have declared these platform workers as essential; nevertheless, labour platforms have not adequately supported them. Consequently, in several countries grassroots movements of platform workers are getting organised to strike and demand better financial and health protection.

Once again, platform work is in the spotlight, and media coverage and social pressure are calling for an opportunity to secure better working conditions for platform workers (Wired 2020a). On 27 March 2020, the US Congress passed the CARES Act, which extended unemployment benefits to self-employed and platform workers under the Pandemic Unemployment Assistance programme. However, the implementation is proving not to be smooth (Wired 2020b) and it is not evident how many platform workers will obtain the subsidy and on what basis (that is, number of hours worked in the last month? Percentage of personal income deriving from platform work?). These difficulties are for policy-makers examples of the importance of knowing the prevalence and characteristics of platform work.

To tackle the challenges that the transformation of labour is imposing on us, a necessary first step is to have sound estimates of platform work. Policy-makers need to know the number of people who spend a significant amount of time and gain substantial income through labour platforms, as well as the type of work they supply, in order to correctly address the policy intervention. Furthermore, it is essential to understand whether platform work constitutes the main source of income or an additional gig on the side of regular employment.

In the next section, I briefly present previous attempts at measurement and then describe the associated challenges.

CHALLENGES FOR MEASURING PLATFORM WORK

A McKinsey Global Institute (2015, p. 1) study claimed that by 2025 'online talent platforms could raise global GDP by up to $2.7 trillion and increase employment by 72 million full-time-equivalent position'. Several measurement attempts have followed this study using different approaches and techniques.

A first approach consists of using information readily available, either from the platforms or other sources, to derive preliminary estimates of the emerging phenomenon. Farrell and Greig (2016) derive the prevalence of platform work in the USA using information from JPMorgan Chase customers (nearly 6 million people) who received income at least once over the 36 months preceding the study from at least one of 42 online platforms. They estimate that, in June 2016, 0.5 per cent of the adult population gained money via labour platforms, and 1.5 per cent over the entire period of the study. Those numbers rise, respectively, to 0.9 per cent and 4.3 per cent if earnings from capital platforms are also taken into account. A study by the European Commission (Nunu et al. 2018) estimates indirectly the number of platform workers in Europe combining data on platform revenues and aggregate data on sectoral employment. They estimate that 1.79 per cent of European workers provide services via online platforms. However, this figure is potentially misleading, as it refers to full-time equivalent, a concept scarcely applicable to platform work.

The second main approach is to estimate the number of platform workers on the basis of dedicated surveys that directly ask respondents whether they provide services via digital labour platforms (Katz and Krueger 2016; Huws et al. 2017; Lepanjuuri et al. 2018; Pesole et al. 2018, Urzì Brancati et al. 2020). Katz and Krueger (2016) estimate that only 0.5 per cent of people aged 18 years and over in the USA undertake platform work as a main job. According to Huws et al. (2017), about 2.9 per cent of the people in their sample of seven European countries earned at least 50 per cent of their income from platform work. Lepanjuuri et al. (2018) estimate that approximately 4.4 per cent of people aged 18 and over in the United Kingdom carried out platform work during the previous 12 months. Pesole et al. (2018) estimate that approximately 2.3 per cent of the respondents in their sample of 14 European Union (EU) member states provide services via platforms as their main job. Building on the latter, Urzì Brancati et al. (2020) extend the country coverage of the analysis to 16 EU member states and find that, in 2019, 1.4 per cent of the working-age population provides services via platforms for more than 20 hours per week and/or earns more than 50 per cent of their income via platforms.

Several National Statistics Offices also introduced specific questions to identify platform workers. However, their measurement approaches differ depending on the objectives and the context where questions were introduced. In 2017, the US Bureau of Labor Statistics (BLS) added four questions to the Contingent Worker Supplement with the aim of measuring platform work. The questions were designed to capture electronically mediated work defined as 'short jobs or tasks that workers find through websites or mobile apps that both connect them with customers and arrange payment for the tasks' (CPS 2018, p. 1). The proportion of platform workers in the USA after data collection was 3.3 per cent of total employment, nearly 6 million people. However, according to the BLS the questions did not work as intended and, after a careful revision, the collected data presented a large number of false positives (that is, incorrectly defined platform workers). Following the revision, the proportion of platform workers has been lowered to 1 per cent of total employment, corresponding to 1.6 million people (CPS 2018). Likewise, the estimates for Canada (Statcan 2017) are particularly low at 0.3 per cent; however, the scope of the question is limited to people who provide services such as 'rides' facilitated by online platforms such as Uber or Lyft.

Finally, the Online Labour Index (OLI), developed by the Oxford Internet Institute 'provides the online gig economy equivalent of conventional labour market statistics; it measures the utilisation of online labour across countries and occupations by tracking the number of projects and tasks posted on platforms in near-real time' (Kässi and Lehdonvirta 2018, p. 1). The OLI tracks the top five fully digital labour platforms and covers online freelancing only. As a consequence, the OLI does not take into account platform workers who provide services on location. Furthermore, the calculation assumes that the same users do not use multiple platforms; if they do, there is a risk of double counting (Kässi and Lehdonvirta 2018).

Despite the numerous sources of data collection, homogenous information on the prevalence of platform work is still lacking. As discussed next, the reasons for this disparity in the estimates are multiple and highlight each specific challenge that measurement entails. First, definition of platform work and its functional characteristics may differ across empirical studies that measure the prevalence of platform work. For instance, many of the studies treat separately capital[1] and labour platforms, as well as distinguish the provision of services and goods. Secondly, the measurement approach used may produce different estimates owing to differences in the choice of the reference period and target population. That is, the results may vary according to the length of the reference period used to measure the prevalence (that is, provision of labour services over the past week, past month or past 12 months). Also, results may vary if the question about provision of labour services via platforms is asked only of a restricted target of people (for example, the self-employed). Finally, the type of task performed and attached working conditions could affect the final estimate of platform workers. That is, platforms may be considered as employers, in a broader sense, and platform workers could also be considered as employees directly hired from the platform company. Understanding the task and working conditions of each worker will allow us to distinguish

---

[1]   Here, 'capital' platforms refer to platforms that operate in the sharing economy where platforms enable peers to share or rent out under-utilised assets, such as accommodation (Airbnb) or a car (BlaBlaCar).

between a strict definition of platform workers (those who are not directly hired by the platforms, but whose working conditions are nonetheless still dictated by the platforms) and workers in the platform economy. In principle, both types of information may be relevant for policy purposes, although employees of the platform conceptually should not be considered as platform workers.

Functional Characteristics of Platforms

Whether platforms serve as mere intermediary or play a more active role in the organisation of work has deep consequences both for correctly assessing the employment status of platform workers and for defining the relevant characteristics from a statistical perspective. Indeed, a crucial step for correct measuring consists in the conceptualisation of the attributes of platform work. Several attributes should be taken into account, including: the degree of platform control and intervention in the transaction; the proportion of labour service versus capital rent involved in the transaction; the inclusion of selling goods to generate profits; and, finally, the inclusion of volunteer and unpaid work. Moreover, these conceptual difficulties increase when attempting to find a one-size-fits-all definition for platform work. A good start will be to acknowledge that different purposes may need different definitions. The ideal solution will be to collect as much as information as possible in order to combine them to accomplish different goals. However, such an extensive ad hoc data collection normally represents a big burden, at organisational and economic levels, making it difficult to measure all the different attributes in a single statistical source. In addition, platform work is still too rare a phenomenon among the general population to justify the effort.

The object of measuring is the labour service embedded in the digital transaction. However, if this is a straightforward concept for platforms mediating online labour services (for example, Freelancers, Upwork and AmazonMechanicalTurk) or on-location labour services (for example, Deliveroo, TaskRabbit and Uber), what happens when labour transactions take place on platforms whose primary target is not the supply of labour services? In theory, independently from the nature of the platform, either an e-commerce platform (for example, e-Bay and Etsy), social media (for example, YouTube and Instagram) or a capital platform (for example, Airbnb and Turo), the provision of labour service should be identified as platform work. However, it is not always possible to detect immediately if a contribution from a user constitutes work. To operationalise the definition of platform work, the transactions should satisfy additional criteria. In detail, (1) they should not include income generated by capital rents or other forms, and (2) the matching and the payment must be coordinated by the platforms through a website or an app. Another criterion to facilitate measurement will be to distinguish between different types of platform work, namely, paid work and volunteer and unpaid work. Unpaid and volunteer work, even on digital platforms, present different characteristics from paid labour services and may need a dedicated statistical methodology. Although I recognise the importance and the value of this information for policy purpose, I focus the discussion only on digital labour paid work.

This approach in principle includes a broad range of activities or types of platforms, comprising both goods and services, but also excludes any transactions that do not involve monetary exchange. The primitive collaborative economy is excluded (that is, home-swapping, couch-surfing, wiki platforms and non-equity crowdfunding) as well as other forms of unpaid work (that is, competitive programming platforms). In addition, this approach requires a level of platform intervention that goes a further than mere matching, and excludes job board sites and similar (for example, monster.com and LinkedIn) (Meijerink and Keegan 2019). Still, in practice many surveys have been targeted in order to exclude some sectors or to limit the analysis only to some type of platforms. For example, Italy and the USA target services only, excluding accommodation. Canada, Denmark and Finland include accommodation services, while China and Switzerland extend it to both goods and services.

The Measurement Approach

A second issue is to choose which measurement approach to follow, together with a wording that does not generate confusion with other forms of remote working (for example, teleworking). The two most common approaches are the income-based approach and the job-based approach. The former asks respondents if they earned any income from a list of chosen online activities. The advantage of this approach is that, from the start, it makes it possible to distinguish between the different types of platforms through which workers have supplied their services. In this instance, in order to understand whether the user's contribution in non-labour platforms (that is, e-commerce or capital platforms) constitutes platform work, additional information on the nature of the contribution should be collected. For e-commerce, the additional attribute consists mostly in the intention of buying and selling for profit; for capital platforms, in addition to the rent generated by the capital asset (that is, the rent from Airbnb), additional labour services must be supplied (for example, cleaning the room, preparing breakfast or giving guided tours).

As much as the income approach allows for a large spectrum of activities, it also makes less immediate the need to identify platform workers, as additional questions have to be asked in order to assess the labour input. A satisfactory compromise could be to specify the platforms' options to allow for a quick screen of 'for certain' platform workers with respect to the most blurred cases.

Table 2.1   Screening question COLLEEM questionnaire

Have you ever gained income from any of the following online source?	Yes	No
Selling products or your own possessions on online marketplaces *(e.g. Etsy, eBay and others)*	( )	( )
Renting out accommodation on online platforms *(e.g. Airbnb, Sharedesk, Nestpick and others)*	( )	( )
Leasing out goods on online platforms *(e.g., Turo, PeerRenters and others)*	( )	( )
Crowdfunding or lending money on peer-to-peer lending platforms *(e.g. Kickstarter, Indiegogo, Zopa, Prosper, Kiva and others)*	( )	( )
Providing services via online platforms, where you and the client are matched digitally, payment is conducted digitally via the platform, and **work is location-independent, web-based** *(e.g. Upwork, Freelancer, Timeetc, Clickworker, PeoplePerHour and others)*	( )	( )
Providing services via online platforms where you and the client are matched digitally, and the payment is conducted digitally via the platform, but **work is performed on location** *(e.g. Uber, Deliveroo, Handy, TaskRabbit, MyBuilder and others)*	( )	( )

Source: COLLEEM questionnaire, doi:10.2760/742789.

As an example, Table 2.1 reports the screening question used by Pesole et al. (2018) and Urzì Brancati et al. (2020) in the COLLEEM survey developed by the European Commission's Joint Research Centre (JRC). The screening question requests information about the general online earnings of respondents. The authors of this survey decided on a strict definition of platform work, reflecting the main destination of the platforms. Only respondents who reply positively to the final two options are accounted for as platform workers, maintaining the distinction between fully digital activities and on-location activities.

The second option is the job-based approach. This is used in traditional labour markets by the Labour Force Survey (LFS) and others. The respondent is asked whether in a specific reference period (for example, last week) he or she has supplied at least 1 hour of paid work. If not, they are also asked whether they had a job or business from which they were temporarily absent because of illness, holiday, industrial dispute or education and training. Usually, the reference period for LFS is 'last week'. This has an important implication, namely, the regularity of provision of labour service. Although in traditional labour markets the labour provision takes place within structured jobs that present in most of the cases stability and regularity during the supply of labour, platform work generally is reduced to the accomplishment of a single task with no guarantee of continuity. This potentially could lead to seriously biased estimates, depending on the reference period chosen and the minimum amount of time worked. As an illustration, let us consider a reference period of a week and a minimum of 1 hour of work. Given that platform work is volatile and irregular, and subject to seasonality, the risk of underestimation is very high. Moreover, the concept of 1 hour of work could be misleading. Many tasks performed via platforms last less than 1 hour, as is the case particularly for microtasks, and, when considering delivery or transportation, technically their paid working time is only that related to the provision of the service (again a ride or a drive are usually less than 1 hour) excluding unpaid waiting time. On the contrary, extending the reference period to a longer period (that is, 'last month' or longer) with no limit on the minimum time of work supplied would increase the risk of inflated estimates and that included people who tried platform work once. This has two important corollaries for the purposes of measuring platform work. First, the identification of platform work as a form of employment requires assessing not only whether someone has done some work via platforms during a particular reference period, but also the regularity, intensity and significance of that work. Second, any measure of the attributes of the work performed or the type of labour relationship must be made at the level of the specific tasks (Pesole et al. 2019).

In brief, assuming that platform work could be measured as traditional employment presents several limitations. The unstructured nature of platform work and the alternating of paid working time and unpaid waiting or searching time during a unique working shift makes it challenging for policy-makers and statisticians to find a general rule to apply. Another issue is to identify a common wording to phrase the questions in a way that is understandable for the respondents. The BLS experience showed how, following a review of the data collected, the majority of the identified cases of platform work had to be recoded owing to a general misunderstanding of the questions, lowering the initial estimates from 3.3 per cent to 1 per cent of the adult population.

In order to simplify the theoretical development of the many aspects that should be taken into account when designing a potential questionnaire for platform work, and just for the sake of illustration, I compare the screening question used by COLLEEM and that used by the BLS in the US Contingent Worker Supplement[2] (Figure 2.1).

---

[2] The comparison is not between the quality or the validity of the questions and the surveys. The two surveys have very different scope and magnitude and so cannot be remotely compared. It is just an illustration of two approaches.

PESEM   Some people find short, IN-PERSON tasks or jobs through companies that connect
        them directly with customers using a website or mobile app. These companies also
        coordinate payment for the service through the app or website.

        For example, using your own car to drive people from one place to another, delivering
        something, or doing someone's household tasks or errands./ [READ ONLY IF
        NECESSARY: For example, using your own car to drive people from one place to
        another, delivering something, or doing someone's household tasks or errands.]

        Does this describe ANY work (you/NAME) did LAST WEEK?

                (1) Yes   [Go to PESEMWJ]
                (2) No   [Go to PESEMWJ]
        [Blind] (D) Don't Know   [Go to PESEMWJ]
        [Blind] (R) Refused   [Go to PESEMWJ]

Figure 2.1   Screening question in the US Contingent Worker Supplement

The approach used by the BLS is to define the type of the arrangements in the
question; I refer to it as a tool-based approach. This is not dissimilar to the
final two options in the COLLEEM questionnaire screening question (Table
2.1), where the type (that is, matched digitally) and the tools (that is, online
platforms, mobile app, and so on) needed for the work arrangements are noted
in detail. The first difference is that COLLEEM follows an income approach,
while the BLS asks for provision of work. A second and important difference
between the two surveys is the inclusion of examples in the questions. The
COLLEEM questionnaire includes example of names of platforms; the BLS
does not. This decision is not as neutral as it could appear. On the one hand,
examples may help in clarifying difficult concepts; on the other, they may
bias the answer (not all platforms could be listed) and create problems of
consistency across countries and over time. A final important difference is
the reference period: COLLEEM asks for an indefinite period of time (ever
gained), while the BLS reference period is 'last week'. The advantage of
using 'last week' as reference period is to enable comparison with a traditional
labour survey. Nonetheless, platform work is characterised by short run performances and is highly subject to seasonality implying the need for a longer
time horizon.

   Pesole et al. (2019) propose some practical principles for the measurement
of platform work that could help in balancing the different statistical needs.
After determining whether the respondent supplied any labour services via
digital labour platforms, the second step they suggest is to qualify the provision
of labour services via digital labour platforms in at least the following respects:

1.   Locus (online or in-person provision).

2. Regularity (how often) over a specific reference period (for example, last month).
3. Time allocated (working hours) in the reference period.
4. Income generated (in monetary terms and as a share of total personal income).

Following this structure, the screening question about the general provision of services could keep a longer time horizon (for example, during the 'last 12 months') compensated by a shorter reference for the question to attest regularity, working hours and generated income (for example, 'last month'). Using information on the regularity, time allocated and income generated via digital labour platforms, thresholds can be set to classify platform employment as:

1. Main (equivalent to a regular job: regular, enough hours, enough income).
2. Secondary (regular but less than enough hours and income).
3. Sporadic (infrequent and inconsequential in relation to time or income).

This modularity enables the different requests from various policy needs to be met. If it is important for policy-makers to understand the full size of the phenomenon, equally the scope for intervention changes dramatically according to the dependence of the workers from the platform. That is, a reference period of 12 months allows for also capturing those workers who only provide services sporadically, who most likely, however, will not be the primary target of policy intervention. A shorter reference period (for example, 'last month' or 'last week') is more suitable for asking questions about the characteristics of the digital labour platform (that is, working hours, number of tasks performed and working conditions, and so on).

   An additional issue to consider when analysing the working conditions of platform work is the variety of the type of tasks available on platforms and the possibility that platform workers engage in more than one type of activity. This implies that, unlike a traditional labour market survey, where normally the respondent has a clear reference for his or her working conditions (that is, the main job), a similar concept is difficult to translate to platforms when they perform more than one type of activity.

The Task-based Approach

A central aspect to regulate platform work is to learn about the working conditions of platform workers. In order to do so, the analysis should be specific to the task performed. Indeed, online freelancers and riders face different challenges and present specific working conditions. In addition, it is very common for platform workers to perform more than one type of tasks and to supply labour via different platforms, which add another layer of complexity when analysing the data collected. Creating a taxonomy of tasks for digital labour platforms comes with practical difficulties. The use of a tasks framework that also applies to regular work (for instance, Bisello et al. 2019) can be useful for comparing the tasks carried out within platform work and regular work. However, these frameworks are too general and, although they may be used as guidelines, they do not allow for unambiguously mapping tasks to specific occupations. In addition, the content and the skill requirements change within occupations, which makes any attempt to reduce this complexity in a fixed classification of tasks potentially problematic. Nonetheless, practical help could come from the taxonomy used by the same platforms bearing in mind, however, that some degrees of inconsistency remain. Broadly, task types can be differentiated according to the locus of provision (online or on location), the skill level and complexity (professional versus non-professional), and the scale (large versus small tasks).

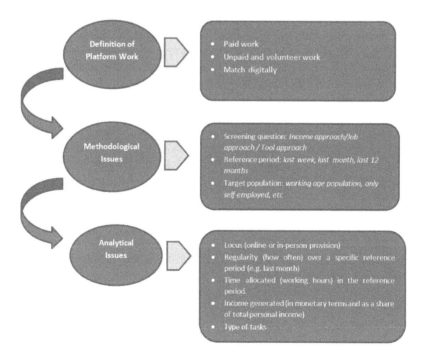

Figure 2.2   Digital platform work and measurement criteria

Figure 2.2 summarises schematically the challenges of measuring platform work discussed in this section. It starts with the selection of the potential attributes that may contribute to the definition of platform work. Then it moves to the definition of the methodological aspects, and ends with the inclusion of the characteristics to describe digital platform work and classify the different types of platform workers (that is, main, marginal, sporadic and so on).

SIZE AND NATURE OF PLATFORM WORK IN EUROPE: EVIDENCE FROM THE COLLEEM SURVEY

In this section, using data from the COLLEEM survey[3] developed by the European Commission's JRC, I present a practical implementation of the issues discussed in the previous section and some facts about platform work in Europe. The JRC COLLEEM survey is an online panel survey covering 15 EU member states[4] plus the UK, and gathers a total of 38 878 responses from Internet users aged between 16 and 74 years old. A common weakness of online surveys is that respondents are drawn from a commercial online panel, which might be a source of bias itself since participants could be more engaged in online work than are others among the general population. For the same reason, also within the identified platform workers, those who supply labour on location might be underrepresented (see Urzì Brancati et al. 2020). Although with some limitations, COLLEEM is currently the major source of information available on comparative data on platform work in Europe.

The big question is about the size of platform work. This question does not have a unique answer. It depends on the conceptual definition of what constitutes platform work, what constitutes a digital labour platform, the methodological approach used to collect the data and the reference period.

The COLLEEM survey contains a direct measure of labour service provision via platforms. It asks whether the respondent has ever gained income from different online sources, among which there are two corresponding to labour service platforms (final two rows in Table 2.1). That is, COLLEEM estimates exclude any labour supplied via capital platforms (for example, Airbnb) or any additional income deriving from e-commerce, crowdfunding and similar activities. In addition, the question asks if the respondent has ever gained any income, hence with no restriction over a specific reference period. The prevalence of platform work in this broader definition is around 11 per cent of the working-age population for the 16 countries covered, ranging between 18 per cent in Spain and 6 per cent in the Czech Republic.

---

[3]   The COLLEEM survey was carried out for the first time in 2017 and repeated in 2018.
[4]   The countries covered are Croatia, Czech Republic, Finland, France, Germany, Hungary, Ireland, Italy, Lithuania, the Netherlands, Portugal, Spain, Sweden, Slovakia, Romania, and the United Kingdom (still an EU member state at the time of data collection).

However, these figures give an indication of how spread the phenomenon is in the countries analysed, but it is not yet definable as an indication of how many people spend a consistent amount of time providing services via platforms or make a living out of platform work. Figure 2.3 shows the different estimates when frequency (labour services are provided via platform at least monthly), regularity (number of hours worked) and income are taken into account.

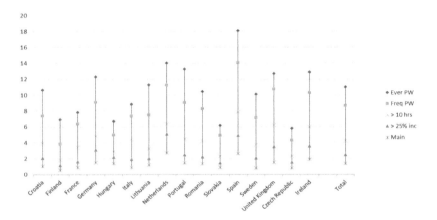

Figure 2.3   *Estimates of platform work adjusted by frequency and income*

According to COLLEEM data, about 11 per cent of the working population in the 16 countries has tried platform work at least once: these are denoted in Figure 2.3 as ever platform workers (Ever PW). Frequent platform workers amount to 8.6 per cent and are platform workers who have provided labour services at least monthly (Freq PW). When the number of hours worked are taken into account, the percentage drops to about 4 per cent for those who worked via platforms for more than 10 hours in the last month (>10hrs) and, if additional income is taken into account, the share of platform workers who made at least 25 per cent of their income from platform work goes down to 2.5 per cent. Finally, main platform workers (Main) are defined as those who provide labour services via platforms at least monthly, and work on platforms at least 20 hours a week or get at least 50 per cent of their income via platforms. They account for 1.4 per cent of the working population in the 16 countries considered.

Who Are the Platform Workers and What Do They Do?

The narrative about platform work, in particular when it started, was about young people seizing the opportunity for some extra cash or testing their expertise on the digital market. Conversely, the data shows that the typical European platform worker is 35 years old, male, educated to degree level and with family commitments. Over 50 per cent of platform workers have children and the majority lives in households of three or four people. Men are more numerous among platform workers (65 per cent), although comparing COLLEEM data in 2017 and 2018 the participation of young women has increased. In 2018, COLLEEM also collected data on foreign-born workers. The data in Figure 2.4 suggest that a much higher proportion of foreign-born workers provide services via digital labour platforms than do native workers. However, this information does not have a unique interpretation. On the one hand, the large presence of foreign-born platform workers may suggest that work on digital labour platforms is not particularly attractive, since several studies have demonstrated how foreign-born workers tend to be employed in lower-quality jobs and be overqualified (OECD/EU 2018). On the other hand, the results may be an indication of inclusivity.

	Offline workers	Marginal	Secondary	Main PW	Number of platform workers
Ireland	28,3%	39,3%	36,8%	50,6%	322
Portugal	11,2%	13,7%	16,2%	8,8%	382
Spain	10,2%	18,7%	16,0%	13,7%	477
United Kingdom	8,9%	21,4%	18,0%	29,2%	297
Sweden	8,6%	24,9%	27,7%	26,8%	210
Netherlands	8,1%	24,1%	16,6%	15,7%	314
Croatia	7,6%	15,0%	14,8%	25,5%	262
Czech Republic	5,4%	17,5%	18,0%	10,7%	143
Germany	4,6%	10,3%	11,2%	6,5%	252
France	4,1%	23,0%	15,1%	5,1%	213
Lithuania	3,1%	0,7%	4,6%	0,7%	213
Slovakia	3,0%	7,0%	4,7%	0,0%	162
Finland	2,9%	25,4%	37,8%	36,7%	99
Italy	2,5%	9,5%	9,0%	8,3%	349
Hungary	2,2%	5,7%	9,5%	0,8%	163
Romania	0,9%	4,0%	0,0%	0,0%	335
Total	6,0%	16,3%	14,4%	13,3%	4.193

Figure 2.4   Foreign-born workers by country and intensity of platform work

To understand better the working conditions of platform workers, it is important to know what they do. Building on Kässi and Lehdonvirta's (2018) definition of fully digital tasks by, COLLEEM elicits information on 10 different types of task, combining locus of provision and skill level. The 10 task types defined by the COLLEEM survey are the following:

1. Online clerical and data-entry tasks (for example, customer services, data entry, transcription and similar).

2. Online professional services (for example, accounting, legal, project management and similar).
3. Online creative and multimedia work (for example, animation, graphic design, photograph editing and similar).
4. Online sales and marketing support work (for example, lead generation, posting advertisements, social media management, search engine optimisation and similar).
5. Online software development and technology work (for example, data science, game development, mobile development and similar).
6. Online writing and translation work (for example, article writing, copywriting, proofreading, translation and similar).
7. Online micro tasks (for example, object classification, tagging, content review, website feedback and similar).
8. Interactive services (for example, language teaching, interactive online lessons, interactive consultations and similar).
9. Transportation and delivery services (for example, driving, food delivery, moving services and similar).
10. On-location services (for example, housekeeping, beauty services, on-location photography services and similar).

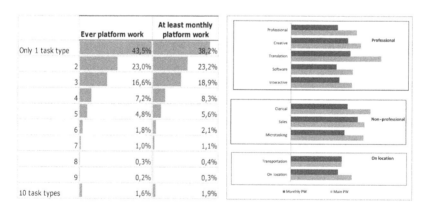

Figure 2.5   Number and distribution of types of tasks

Two elements stand out of platform work in respect of other forms of labour provision. First, platform workers often provide multiple types of services; second, formal education is less important than it is in the traditional labour market (Herrmann et al. 2019). Figures 2.5 reports, on the left-hand side, the number of different task types they do via platforms (not the number of gigs performed), and, on the right-hand side, the distribution of the different types. That is, when a platform worker says that he or she has done more than one type of task, it means, for instance, they did a microtask and a creative task.

The percentage of platform workers providing more than one type of task increases with the intensity of work via platform (left-hand side picture in Figure 2.5). The distribution according to the type of tasks shows a prevalence of platform workers in online activities (that is, professional and non-professional). Urzì Brancati et al. (2020) argue that since COLLEEM is an online survey, platform workers carrying out online tasks may be over-represented in respect of those who carry out on-location services. Furthermore, while online services could be done virtually from everywhere, the provision of on-location services is conditional to the presence of the platform in the territory. This is shown by the map in Figure 2.6. Even if for both online and on-location services the map shows a higher concentration in big cities and metropolitan areas, evidently this effect is stronger for on-location services.

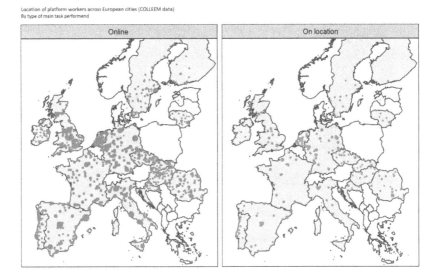

Figure 2.6   Online and on-location platform workers

However, this picture could change quickly, if considering that in only one year Deliveroo went from covering 34 to 154 Italian cities and planned to extend to over 200 by the beginning of 2020.

Other interesting information is to what extent platform workers offer their services in more than one platform (multihoming). According to COLLEEM data, 40 per cent of platform workers[5] use more than one platform, 52 per cent use only one platform and the remaining 8 per cent prefer not to answer. Looking at the data more in detail, those in professional services are most likely to provide services over more than one platform (ranging from 40 per cent in translation services up to 47 per cent in interactive services). However, among platform workers providing on-location and transport and delivery services only 31 per cent and 36 per cent, respectively, provide services in more than one platform.

Figure 2.7 shows how task types are distributed across countries. That is, it shows the relative percentage of platform workers in each specific task by country. In general, for most of the countries, the data do not suggest a particular specification of one type of task over the other. However, some countries report a higher concentration in specific activities. This is the case in Lithuania for clerical and professional tasks, the Czech Republic for transport and on location, and Finland for microtasks, although the percentage of platform workers in Finland is low. Less distinguished is the predominance of software tasks in the United Kingdom and Ireland.

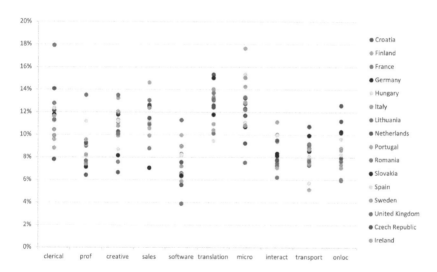

Figure 2.7   Task types distributed across countries

---

[5]   These data refer to platform workers who only perform one type of task, as those performing more than one type of tasks (that is, delivery and microtask) use more than one platform by default.

Hours, Income and Working Conditions

The limitations exposed so far on the measurement of platform work become more evident when trying to assess work and employment conditions. The COLLEEM survey collects some data that could be used to sketch a broad picture of platform workers' working condition. However, as its authors suggest, these figures are problematic and should be treated with caution. One of the questions that COLLEEM asks is on what basis they get paid. About 69 per cent of platform workers report they get paid based on the task performed, which implies that many of them must undertake a substantial share of unpaid work (searching online, waiting time for assignment, and so on) in order to obtain paid work.

The COLLEEM survey also asks about the time spent and income earned in relation to the last task performed. The results reported, however, suggest that the question was misunderstood as, for example, within the same type of task the duration reported for the completion of a single task (that is, gig) varies between 30 minutes and 6 hours or, in some instances, few days. On average, however, those respondents who (possibly) interpreted the duration of last task correctly (for example, duration of the last ride) reported the average duration of the task to be about 30 minutes. In other instances, when respondents interpreted the question as the continuous spell of time they were logged into the app (that is, as a work shift) the average work shift was about 7 hours and, finally, the average number of days to complete a task was 11 days. The COLLEEM survey also elicits information about how much respondents get paid for the last task. The question is plagued by the same level of misunderstanding as that on duration. However, Urzì Brancati et al. (2020), by restricting the sample to a pool of consistent respondents, provides some information on the average payments for the seven top platforms in the dataset. Table 2.2 reports the median payments for selected platforms in the COLLEEM dataset. Two clarifications must be made about the data. First, the values reported are the average values for all 16 countries. The price of an Uber ride in London will probably be different from the price in Prague. Secondly, for those platforms offering homogenous types of tasks, the reported median values do not vary much, whereas on platforms such as Freelancer or PeoplePerHour the variation among task type could be substantial. As the survey's authors report,

> the lowest median value in Freelancer is €5 per hour for micro-task jobs and €22 per hour for interactive services. There are no other official sources we could use to compare our findings with; however, the Fair Crowd Work website collects information on the median payment in some selected platforms. The only two platforms that overlap are Clickworker and Upwork for which they report respectively a median value of €2.92 and €12.91. (Urzì Brancati et al. 2020, p. 34)

Table 2.2   *Median payment by platforms (€)*

Platforms	Pay per hour (median)	Pay per task (median)	Total observations
Clickworker		2.3	63
Deliveroo	11.5	30	27
Fiverr		16	42
Freelancer	11.0	67	793
PeoplePerHour	25.0	64	62
Uber	8.5	67	181
Upwork	8.0	41	67

In a consequential effort of data cleaning and analysis, Urzì Brancati et al. (2020) also report an estimate of pay per hour for the different types of task. According to their findings, the highest hourly paid tasks are software coding (€23) and interactive services (€16). Clerical, professional and sales tasks are all around €14 per hour, on-location services and transport vary between €9 and €12 and, finally, the least paid are microtask at about €6 per hour. However, as the authors warn, these numbers should be treated as indicative given the restricted number of observations.

From a policy perspective, an important element to understand is whether platform work is the only source of income and how many platform workers are covered by other forms of social protection. Following the findings in the literature (Huws at al. 2017; Eurofound 2018; Pesole et al. 2018; Urzì Brancati et al. 2020), the dominant economic status reported by platform workers is dependent employment. For instance, according to Urzì Brancati et al. (2020, p. 52), among main platform workers 25 per cent report to be employees and among secondary platform workers the share increases to 36 per cent. This suggests that most platform workers have additional regular jobs and use digital labour platforms as a secondary source of income. On the one hand, this may raise concerns that people in regular employment need to find alternative and additional sources of income; on the other, that platform work is in most cases a secondary activity makes its conditions less consequential for the risk of exclusion from social protection for platform workers. Even for those workers covered by traditional labour protection institutions, the additional burden in relation to hours may result in risk for their well-being. Indeed, platform workers tend to work longer hours (both inside and outside the platform) and frequently these hours are during atypical or unsocial working times. In particular, more than two-thirds of all platform workers provide their services via platforms at weekends, and nearly the same amount provide their services at night (Urzì Brancati et al. 2020).

Finally, COLLEEM includes some questions intended to measure the specific psychosocial conditions of platform work. The answers indicate that, although the majority of platform workers value flexibility associated with platform work, for half of the workers platform work creates a degree of stress, is often perceived as monotonous and, in most instances, is subject to constant monitoring by the platform.

CONCLUSIONS AND FUTURE RESEARCH

Measurement of platform work is still a challenge for policy-makers and researchers. The issues presented in this chapter contribute to the debate, offering some practical examples for reflection on how to tackle the various aspects of such a complex task. In order for a phenomenon to be understood, regulated and coordinated, it must first be measured. Indeed, measuring platform work will help us to shed light on accessory elements still unrevealed.

In this chapter, I discussed the advantages and disadvantages of different approaches focusing on the methodological challenges of collecting data through an ad hoc survey. The dimensions analysed range from the challenge of finding an encompassing definition of platform work that will suit policy purposes and that respondents will clearly understand, to the definition of the attributes that will shape the final metrics for the estimate of platform work prevalence.

The chapter presents the main components for the delineation of a statistical framework, addressing both conceptual and operational challenges. Among these, a first element is the distinction between provision of services or goods, or capital versus labour platforms. While in a traditional labour market it is irrelevant for measurement purposes whether the outcome of labour is a good or a service, in digital labour platforms, this distinction may determine sizeable differences in the estimate of platform work prevalence. Indeed, selling goods or services, such as renting out an apartment, may be excluded on the ground that the relative labour input in the transaction is low.

A second element is the design choices of the methodological approach. Throughout the chapter, I discuss how estimates might be inflated or deflated according to the chosen length of the reference period or the chosen approach for the screening question to identify platform workers (for example, income, job or tool-based approach). Finally, an essential aspect is the importance of integrating a task-based approach to understand thoroughly how work content and work organisation are affected by platforms' intervention. The reorganisation of work that platforms entail has consequences that go beyond the deconstruction and circumvention of traditional labour market institutions, by also affecting measurement and making it more difficult to provide evidence-based support to policy-makers.

In this chapter, and in addition to methodological issues, I also present some findings from a new source of data, the COLLEEM dataset. The findings presented hint at platform work as an alternative source of income for workers in distress, to whom jobs do not guarantee decent living conditions and with an ambiguous employment status. The ambiguity of their employment status contributes to increasing the plethora of workers that do not see their basic rights recognised. Minimum wage, paid holidays, maternity leave and sick leave, as well as right to freedom of association and collective bargaining, are policy instruments that already exist and simply do not cover all workers. In the past 30 years, the flexibilisation of labour markets has introduced the generally accepted idea that jobs should not guarantee stability. On the contrary, they should follow the productions needs and be regulated by market demand. Digital labour platforms reinforce this belief and facilitate an even more extreme market-based redistribution of the labour force, introducing new ways of coordinating labour and workers by means of algorithms. The use of algorithms to match, manage and control a multitude of workers – potentially everywhere in the world – has deep consequences on how the relationship between technology and work is evolving, and affecting platform workers' working conditions in relation to access to tasks, remuneration and working-time flexibility. Moreover, the digital management practices originated by platforms are already expanding to traditional markets. This is the case, for example, for ratings, monitoring and surveillance techniques, with consequences for workers' well-being, safety, privacy, dignity and ability to work, and in particular their autonomy. Digital technologies can lead to new forms of precariousness and low job quality, raising concerns about sustainable growth and fair working conditions for workers. However, technological artefacts cannot act independently of human action, which leave scope for using technology to improve working conditions and resolve unfair situations. Future research should aim at shedding light on how current algorithmic management works and how a new techno-paradigm based on transparency, equity of treatment and justice could be implemented. Furthermore, most of the focus on platform work has concentrated on its expansion in developed countries, although research from both the International Labour Organization and the Oxford Internet Institute shows how the vast bulk of demand comes from high-income countries and the majority of work is carried out in lower-income countries. A global effort should be undertaken to agree on a common definition of the phenomenon and to adopt measures that guarantee the dignity of platform workers worldwide.

REFERENCES

AppJobs Institute (2020), 'The future of work', AppJobs Institute, accessed 18 March 2021 at https://irp-cdn.multiscreensite.com/ec5bfac6/files/uploaded/AppJobs %20Institute%20Future%20of%20Work%20Report%202020.pdf.

Bisello, M., E. Peruffo, E. Fernandez-Macias and R. Rinaldi (2019), 'How computerisation is transforming jobs: evidence from the European Working Conditions Survey', *JRC Working Papers on Labour, Education and Technology 2019-02*, Joint Research Centre, Seville.

Cedefop (2020), 'Developing and matching skills in the online platform economy: insights from Cedefop's CrowdLearn study', Luxembourg: Office for Official Publications.

Cook, C., R. Diamond, J. Hall, J.A. List and P. Oyer (2018), 'The gender earnings gap in the gig economy: evidence from over a million rideshare drivers, NBER Working Paper No. 24732, National Bureau of Economic Research, Cambridge, MA.

Current Population Survey staff (CPS) (2018), 'Electronically mediated work: new questions in the Contingent Worker Supplement', *Monthly Labor Review*, September, US Bureau of Labor Statistics, Washington, DC, doi:10.21916/mlr.2018.24.

Duggan, J., U. Sherman, R. Carbery and A. McDonnell (2020), 'Algorithmic management and app-work in the gig economy: a research agenda for employment relations and HRM', *Human Resource Management Journal*, **30** (1), 114–32.

Eurofound (2018), *Employment and Working Conditions of Selected Types of Platform Work*, Luxembourg: Publications Office of the European Union.

European Centre for the Development of Vocational Training (Cedefop) (2020), *Developing and Matching Skills in the Online Platform Economy: Insights from Cedefop's CrowdLearn Study*, Luxembourg: Office for Official Publications, accessed 20 April 2021 at http://data.europa.eu/doi/10.2801/588297.

Farrell, D. and F. Greig (2016), 'Paychecks, paydays, and the online platform economy: big data on income volatility', JPMorgan Chase, accessed 18 March 2021 at https://www.jpmorganchase.com/content/dam/jpmc/jpmorgan-chase-and-co/institute/pdf/jpmc-institute-volatility-2-report.pdf.

Frenken, K. and J. Schor (2017), 'Putting the sharing economy into perspective', *Environmental Innovation and Societal Transitions*, **23** (June), 3–10.

Herrmann, A., P. Zaal, M. Chappin and B. Schemmann (2019), 'Does education still matter in online labor markets?', in I. Maurer, A. Oberg and D. Wruk (eds), *Perspectives on the Sharing Economy*, Newcastle upon Tyne: Cambridge Scholars.

Huws, U., N.H. Spencer, D.S. Syrdal and K. Holts (2017), *Work in the European Gig Economy: Research Results from the UK, Sweden, Germany, Austria, the Netherlands, Switzerland and Italy*, Brussels: Foundation for European Progressive Studies, UNI Europa and the University of Hertfordshire.

Kässi, O. and V. Lehdonvirta (2018), 'Online labour index: measuring the online gig economy for policy and research', *Technological Forecasting and Social Change*, **137** (C), 241–8.

Katz L.F. and A.B. Krueger (2016), 'The rise and nature of alternative work arrangements in the United States, 1995-2015', NBER Working Paper No. 22667, National Bureau of Economic Research, Cambridge, MA.

Kenney, M. and J. Zysman (2016), 'The rise of the platform economy', *Issues in Science and Technology*, **32** (3), 61–9.

Kuhn, K.M. and A. Maleki (2017), 'Micro-entrepreneurs, dependent contractors, and instaserfs: understanding online labor platform workforces', *Academy of Management Perspectives*, **31** (3), 183–200.

Lepanjuuri, K., R. Wishart and P. Cornick (2018), 'The characteristics of those in the gig economy', Research Paper: 2018 no. 2, Department for Business, Energy and Industrial Strategy, London.

McKinsey Global Institute (2015), 'A labor market that works: connecting talent with opportunity in the digital age', McKinsey Global Institute, Chicago, IL.

Meijerink, J. and A. Keegan (2019), 'Conceptualizing human resource management in the gig economy', *Journal of Managerial Psychology*, **34** (4), 214–32.

Nunu, M., R. Nausedaite, K. Eljas-Taal, K. Svatikova and L. Porsch (2018), *Study to Monitor the Economic Development of the Collaborative Economy at Sector Level in the 28 EU Member States*, Brussels: European Commission Directorate-General for Internal Market, Industry, Entrepreneurship and SMEs.

Organisation for Economic Co-operation and Development and European Union (OECD/EU) (2018), *Settling in 2018: Indicators of Immigrant Integration*, Paris: OECD and Brussels: European Union.

Pesole, A., E. Fernández-Macías, C. Urzí Brancati and E. Gómez Herrera (2019), *How to Quantify What Is Not Seen? Two Proposals for Measuring Platform Work*, Seville: European Commission.

Pesole, A., C. Urzì Brancati, E. Fernández-Macías, F. Biagi and I. González Vázquez (2018), *Platform Workers in Europe. Evidence from the COLLEEM Survey*, EUR 29275, Luxembourg: Publications Office of the European Union.

Stanford, J. (2017), 'The resurgence of gig work: historical and theoretical perspectives', *Economic and Labour Relations Review*, **28** (3), 382–401.

Statcan (2017), 'The sharing economy in Canada', accessed 20 April 2021 at https://www150.statcan.gc.ca/n1/en/daily-quotidien/170228/dq170228b-eng.pdf?st =SSvPboyU.

Urzì Brancati, C., A. Pesole and E. Fernández-Macías (2020), *New Evidence on Platform Workers in Europe. Results from the Second COLLEEM Survey*, EUR 29958, Luxembourg: Publications Office of the European Union.

Wired (2020a), 'This pandemic is a "fork in the road" for gig worker benefits', *Wired*, accessed 18 March 2021 at https://www.wired.com/story/gig-worker-benefits-covid -19-pandemic/.

Wired (2020b), 'Gig workers' new unemployment benefits won't come quickly', *Wired*, accessed 18 March 2021 at https://www.wired.com/story/gig-workers -unemployment-benefits-wont-come-quickly/.

3. The past, present and future of gig work

Jim Stanford

INTRODUCTION

There has been a flourishing of new businesses in the past decade offering services through digital platforms that dispatch service providers to perform discrete tasks for end-users. This on-demand or gig business model uses smart-phone technology to receive requests from users (typically via a firm-specific application) and allocate workers to perform the requested work. These businesses function on the basis of an on-demand workforce: workers who are engaged only when a customer wishes to use their services, who provide their own tools and equipment, and who are compensated on a piece-work basis.

The expansion of this business model, and its spread into an increasing range of activities and sectors, has sparked contrasting responses. Some welcome the practice as a sign of independence and empowerment for the workers who perform these gigs. Others fear its consequences for the stability and quality of work. However, both perspectives tend to understand the rise of digital platform businesses as representing something fundamentally new in the organisation of work, and as a trend driven primarily by technology.

This chapter provides a historical perspective on the expansion of digital platform businesses, to better understand the varied forces (not just techno-logical, but also economic and socio-political) that explain its rise. It shows that the major features of gig work have been utilised in employment for centuries, from the earliest days of capitalism and wage labour. These features became less common in industrialised countries during the twentieth century, especially after the Second World War, when more work (but not all) came to be organised on a more stable and permanent basis. Now those practices are becoming more common; the rise of gig work is just one manifestation of the growing precarity of work under neoliberalism. In this light, the rise of the gig economy should be seen as a return to an older, more precarious and exploitive model of employment – not as a creative and progressive innovation. A more complete understanding of the historical antecedents of modern gig practices, and of the full range of forces that drive waves of change in work organisation, can better inform regulatory and political responses to the rise of platform work.

This chapter is organised as follows. The next section discusses the main features of modern digital platform work and the recent expansion of this business mode, and asks whether the practice indeed represents a fundamentally new feature of work and work relationships. The third section briefly considers whether gig work constitutes employment in a legal and practical sense, and confronts claims (favoured by platform entrepreneurs) that gig workers are independent business operators, not employees. It concludes that gig work is indeed employment in the practical sense of that term (even though legal debates will continue): the work is performed for someone else, under their direction and control, and the resulting output is owned by the employer, which pays compensation to those who did the work. The fourth section reviews several historical antecedents for modern gig-work practices, and shows there is long experience with these strategies. The penultimate section considers the broader macroeconomic, political and regulatory forces that also shape employment relationships, and identifies several factors (not just technology) that have contributed to the expansion of gig work. The conclusion in the final section considers the future of gig work, given both the opportunities and the threats these businesses face. If gig work (and other forms of precarious work) reflect a broad composite of economic, political and social determinants, then it is not obvious that these practices will necessarily continue to expand.

NOVEL FEATURES OF PLATFORM-ENABLED GIG WORK?

Digital platform businesses can be divided into two broad categories: those which facilitate the exchange of assets and those which facilitate actual work and production.[1] Platforms which facilitate the sale and purchase of assets do not, for the most part, initiate incremental production. They allow consumers or businesses to sell products into a global marketplace: matching sellers of new or used products with buyers. However, the sale of ready-produced assets does not involve new production (other than the work involved in packaging and shipping those products). Moreover, the growth potential of these businesses seems modest compared with other parts of the gig economy.[2]

---

[1] For more on the categorisation of digital platforms, see Farrell and Grieg (2016), Manyika et al. (2016), Torpay and Hogan (2016) and Valenduc and Vendramin (2016).

[2] Consider, for example, eBay, which was the leading firm in the asset sale and resale category of platform businesses. After an initial wave of excitement surrounding its establishment and early growth, the company has been eclipsed by other platform

Our main focus here is on platforms that engage and organise labour to undertake value-added production.[3] Within this category, an additional distinction can be considered. Some platform businesses, known as crowd work sites, function essentially as digital bulletin boards to advertise labour services. They provide a platform for independent producers to advertise their services to prospective buyers; these producers control their own work, and set their own terms and prices. The platform's role is limited to matching workers with customers, in return for a fee that presumably reflects this intermediation service. The platforms do not organise or oversee production, set prices or otherwise fulfil functions of an employer.

Another class of digital platforms, associated with the largest and most dynamic players in the gig economy, plays a more hands-on role in initiating, managing and financing work and production. These platforms have more potential to transform employment relationships. The best-known early examples of this class of business were in the passenger transportation industry: companies providing taxi-like services organised through digital applications (apps). Uber (founded in 2009) is the largest and most prominent, but many competitors now provide similar services: including Lyft, Via, Ola and Didi. This industry was initially named 'ride-sharing', conveying the impression that drivers were simply sharing their vehicles with paying passengers. That name gave rise to the cheerful and highly misleading term 'sharing economy', to describe the whole set of digital platform businesses. Both these terms have (thankfully) fallen out of use. A better descriptor for this taxi-like business is 'ride-sourcing'.

On one level, ride-sourcing hardly differs from conventional taxi services. The fundamental productive process is identical: a driver is instructed to pick up a passenger at one location, drive them to another and is then paid for the service. However, the work is now arranged through a digital app (whereby customers use smartphones to submit their requests, which are processed centrally and assigned to drivers), instead of previous dispatch methods (such as radio services). However, finding a new way to tell a taxi driver where to

---

businesses that facilitate actual work and production. eBay has spun off its lucrative online payments system (PayPal) and recently sold its advertising business.

[3]   Some platform firms incorporate a mixture of productive work and asset sales or rentals. For example, the online accommodation broker Airbnb was initially designed as a way for property owners to rent (or share) extra rooms or other living space. It now facilitates more commercial short-term rental activity and hotel-like services. There is a component of productive labour involved in this work: large Airbnb operators, for example, must hire workers to clean and manage multiple rental sites (while smaller operators do this work themselves). However, the platform itself does not coordinate this work or manage the workers.

pick up their next passenger hardly constitutes a revolution in work. What Uber and its imitators more significantly changed was the relationship between the drivers and the company. Drivers have been reconstituted as independent contractors who operate their own businesses.[4] They are not assigned particular shifts, but can choose when they want to drive (by logging on to the app and thus becoming available for work).[5] Similar to owner-operators in the taxi industry, they own and maintain their own vehicles, but unlike owner-operators they have no control over the revenue generated by their work. They are told where to pick up their passengers, where to take them and advised on the route to take; payment is calculated and collected through the app (cash payments are generally prohibited).

Unlike taxi drivers, ride-share services cannot be hailed by passengers on the street or at taxi stands; they can only pick up passengers assigned by the app, and the process by which the platform assigns fares to drivers is not transparent. A fair system for allocating fares to drivers is always a main concern in the taxi industry; drivers worry that dispatch systems may favour or disadvantage particular drivers. With digital apps, the criteria on which fares are assigned to specific drivers are opaque and controlled unilaterally by the business; individual drivers have no information to determine whether they are receiving fares more or less frequently than other drivers. Ride-share drivers commonly complain that the algorithms assigning their fares are not based solely on driver proximity to a fare, or taking turns among drivers, but instead are programmed to reward drivers for preferred performance, including being logged in for longer periods of time and receiving higher customer ratings (see Simonite 2015).

Many advantages accrue to the platform firms from this restructuring of employment relationships. The app's control over payment allows the company to capture a large share of revenue (over 25 per cent for Uber, in most markets) from what would otherwise be a low-margin, highly decentralised activity. The costs of capital equipment (for the vehicles) are borne by the drivers, who are also responsible for all ownership, maintenance, fuel, licensing and insurance costs, as well as equipment and data charges for their smartphones. Drivers have no guarantee of the number of fares they receive or their hourly income; they are not paid for waiting between fares, nor for driving to pick

---

[4] Even many traditional taxi drivers are not employees of the fleets they work for. Some are waged workers employed by the owners of the car they drive (and, crucially, its corresponding taxi medallion or permit), some own and operate their own vehicles, and yet others fall into various unique legal categories created to reflect the specific relationships between fleets, vehicle owners and drivers.

[5] Many conventional taxi drivers also have the ability to choose when they drive, so this feature of ride-source work is not new.

up a fare. Thus, labour costs are significantly lower than in conventional taxi services; many studies have found that on a net basis (after expenses) many drivers earn less than legal minimum wages (for surveys of this research see Jacobs and Reich 2020 and Stanford 2018). This has allowed Uber to undercut conventional taxi fares[6] and rapidly expand its market. Uber's capital investment is limited to establishing and operating the app and the dispatch system, marketing the service to customers, and managing (increasingly troublesome) regulatory and legal aspects of the business.[7]

The appeal of this business model has sparked many imitators: both in Uber's initial core industry (passenger transportation) and in other industries also amenable to this dispersed but centrally managed production model. Delivery services (for restaurant food, packages and other products) can be organised in a similar manner – sparking a raft of new businesses (including Uber itself, through its Uber Eats and Uber Freight divisions) to challenge conventional delivery systems. Many platform businesses supply workers for odd jobs and tasks (such as Amazon Mechanical Turk, TaskRabbit and Airtasker); these businesses function like digitised labour hire agencies, with the platform siphoning off a share of revenues paid for the contracted labour. Specialised on-demand platforms have been launched in many other industries, including information technology and programming (such as Toptal or Gitcoin), media (Upwork and Freelancer.com), legal services (UpCounsel and LegalZoom) and social and caring services (Care.com and Pager). Employers in many industries are eager to explore the possibility of transforming employment relationships in similar ways: that is, shifting cost and risk to producers, and evading the costs and obligations normally associated with employment (such as minimum wages, insurance, pensions and social benefits).

Official statistics on gig employment are not available, in part owing to the lack of a precise definition of this work. Nevertheless, a growing body of research attests to its growing importance. Chapter 2 in this volume surveys several estimates of the extent of gig work in European countries, and finds that survey-based estimates of the size of platform-mediated gig work vary widely depending on conceptual definitions and methodological approaches that are used. A recent survey in Australia found that 7 per cent of the workforce had performed some on-demand work in the preceding 12 months (McDonald et al. 2019). A Canadian study based on administrative data estimates that over 8 per cent of workers in 2016 performed digitally mediated gig work that year

---

[6]   Unlike regulated taxi companies, whose fares are generally set by local regulators, Uber is free to unilaterally set its own fare structure.

[7]   Despite these advantages, Uber has never made a positive profit; by 2020, with its business damaged by the COVID-19 pandemic and facing increasingly intrusive regulations in many jurisdictions, its cumulative losses reached close to $20 billion.

(Jeon et al. 2019). Official US data on contingent workers (those who do not expect their jobs to continue) suggest 4 per cent of all employed people are in temporary jobs, but that does not capture all gig workers (Kosanovich 2018). United States Federal Reserve data suggests 5 per cent of adult Americans[8] in 2018 performed some work through online temporary or informal roles, not counting asset-selling platforms such as eBay (Board of Governors of the Federal Reserve System 2019, p. 18).[9] It is evident that gig work is growing in both size and scope, and now constitutes a significant segment of all paid work.

In all of these varied applications, digital platform businesses encompass several common core features:

- On-demand work – people working through platform services are engaged only when their services are directly requested by a customer. There is no guarantee of continuing engagement.
- Piece-work compensation – producers are paid a unit price based on the completion of a defined task. They are not paid a wage or salary for time spent at work.
- Self-provision of capital – workers in platform-based activities are required to provide most or all of their own tools, other capital equipment and place of work. The platform firm's capital investment is limited to the provision of central organising, dispatch and financial systems.
- Intermediation – the platform business places itself between the producer and the end-user paying for their services.[10] The resulting ability to make the market enables the firm to control payments and collect a share of revenue.

The use of the Internet and smartphones must be another defining feature of this new business model, but does the mere use of digital communication and management technologies truly constitute a qualitative shift in business practice? Reliance on digital communication technology hardly makes these businesses unique; indeed, it is hard to find any business in industrial countries, in any industry, that does not use the Internet, smartphones and other digital technologies. Also, we can easily imagine gig-type businesses that could use old-fashioned, pre-digital methods to perform the dispatch,

---

[8]   Measured as a share of total employment (rather than as a share of the adult population), this implies that gig work accounts for 7–8 per cent of all employment.

[9]   This report suggests that participation in non-digital informal or on-demand roles (for example, dog-walking, childcare or selling products at flea markets) is still far more common than online roles.

[10]   See Stewart and Stanford (2017) for more discussion of this intermediate position and its implications.

intermediation and financial functions undertaken by modern Internet-based platforms. For example, imagine a business providing labour-for-hire services using non-digital technologies: receiving telephone calls from prospective clients, assigning jobs to workers on its roster (either by telephone or verbally if the prospective workers are gathered in one place) and collecting payment through normal mailed invoices. Structurally that model is indistinguishable from modern odd-job digital platforms (such as TaskRabbit). This firm would not likely survive in business now, given its slow and labour-intensive management process, but its core operation, and the nature of the relationships it mediates with producers and end-users, do not fundamentally differ from the most high-technology, automated gig platform.[11]

A more novel and consequential application of digital technology in on-demand platforms is the use of algorithmic strategies for allocating, supervising and disciplining workers (discussed in detail in Chapter 6 of this volume). Subjecting workers to the instantaneous, omnipresent and seemingly faceless oversight of digital management systems represents a genuine (and often exploitative) innovation in business practice (Henderson et al. 2018). In this regard, the relevance of new technology is once again linked primarily to its impact on employment relationships, more than on the production process itself. This review of the core features of gig work platforms, none of which seem especially earth-shattering, raises the fundamental question that will be explored in this chapter: is gig work genuinely new, or is it actually very old?

UNDERSTANDING CATEGORIES: GIGS, WORKERS OR EMPLOYEES?

Digital platforms aim to shift the costs and risks associated with traditional employment relationships on to producers, thus enhancing the competitiveness and profitability of the business. This has involved careful and deliberate efforts by these companies to avoid the appearance that they employ the people performing the services that they sell.

For example, Uber describes itself as a technology company (not a transportation company), in the business of providing information services to the drivers who use its app.[12] In this fanciful description, the drivers do not work for Uber: Uber works for the drivers. This is strikingly reminiscent of the long-standing illusion in neoclassical economic theory, that what is

---

[11]   Historical examples of low-technology gig-type businesses are considered elsewhere in this chapter.

[12]   Uber's prospectus for its 2019 initial share offering states 'we are fuelling the future of independent work by providing Drivers with a reliable and flexible earnings opportunity' (Uber Technologies Inc. 2019, p. 159).

conventionally termed 'employment' is really a mutually beneficial market exchange between the owners of two broadly equivalent factors of production: labour and capital.[13] In this story, labour and capital engage in mutual and equal exchange to jointly produce goods and services more efficiently than either could on its own, with the gains then shared between them according to their marginal contributions to output. Neither factor is the boss and they both benefit from the trade. Nobel laureate Paul Samuelson famously put it this way: 'Remember that in a perfectly competitive market, it really does not matter who hires whom' (Samuelson 1957, p. 894). Labour could hire capital, as easily and productively as capital hires labour. In practice, labour does not generally hire capital. A rich heterodox economics literature describes the structural reasons why capital not only hires labour, but also directs and disciplines it – through a labour extraction process that is complex, challenging and conflictual (see, for example, Dow 1993 or Bowles and Gintis 2008). Uber's claims are a modern manifestation of the long-standing and ideological claim that capital helps labour by providing workers with tools to do their jobs more productively (and earn higher incomes as a result).

This attempt by digital platforms to deny they are employers is hardly novel. Consider, for example, the Lehigh Valley Coal Company in Pennsylvania. The company argued in court in 1914 that it did not employ its miners. Instead, it was just an intermediary between supposedly independent miners, and the customers who purchased the coal they produced. Lehigh described itself as providing a matching service that allowed miners to market the coal they extracted (just as Uber pretends it is simply matching drivers with passengers). Among other motives for this far-fetched claim, the company was hoping to avoid liability for workers injured in its mine. The presiding judge rejected this argument, concluding that 'by [the miner] alone is carried on the company's only business; he is their "hand," if any one is'.[14]

Other firms, long before the advent of digital platforms, have also tried to position their workers as contractors, not employees. Another famous case involved the delivery and logistics firm FedEx, which claimed its workforce was a fleet of independent small business owners, who just happened to drive FedEx branded trucks, wore FedEx branded uniforms, and were centrally dispatched and directed by FedEx, at rates and according to conditions determined unilaterally by FedEx. This claim, too, was rejected by the US National

---

[13] Classic references for this neoclassical parable include Marshall (1910) and Robbins (1932). See Spencer (2009) for a critical review of the evolution of neoclassical theories of work and labour.

[14] *Lehigh Valley Coal Co.* v. *Yensavage*, 218 F. 547, 552 (2d Cir. 1914), cited by Jost (2011). The author is indebted to Brian Callaci for this reference.

Labor Relations Board, a full century after the Lehigh case.[15] So the attempt by digital platforms to evade normal obligations and responsibilities of employers is hardly novel; they are simply utilising cutting-edge technology in their effort to accomplish this.

Two other far-fetched arguments are invoked by platform businesses to support their claim that engaged workers are not employees. One argument is that since workers are allowed to start and stop work as they desire, they are not employees. By logging on to and off of the app, drivers can choose when they work, and that flexibility is held to be attractive for workers (especially those trying to combine gig jobs with other work or family responsibilities). However, this apparent flexibility in work schedules is inherently constrained by the nature of the market that these drivers serve: since drivers are not guaranteed any minimum hourly income, they are economically compelled to work when there is strong demand for their services. Working weekday rush hours, evenings and weekends does not represent drivers' free choice so much as the economic reality of their business. Another factor limiting this supposed autonomy is the belief among drivers that refusing to accept fares, and/or logging off the app too often, will result in lower probability of being assigned rides in the future.[16] Therefore, drivers are still subject to censure and punishment for not meeting Uber's expectations about availability. Finally, ability of workers to set their own hours of work is not uncommon in many conventional waged employment situations. There are many jobs in which workers voluntarily bid on variable shifts (a common practice in the hospitality industry), are free to work or not at particular times (particularly for tasks which can be performed by lone individuals, such as deliveries, clerical functions, data entry and processing, and simple assembly), or can set their own schedules and simply tally compensable hours (such as researchers and technical workers, agricultural harvesters and repair or maintenance workers). That ride-share drivers can log off their app at any time hardly proves they are not employees in any pragmatic sense of the term.

Another common argument is that workers own the capital that they use at work, and hence are owners not workers. Again, there are many other examples of workers (including waged employees) who are also required to supply tools, equipment and place of work, including skilled trades, entertainment and arts producers, personal service workers, resource workers and harvesters, home workers in the textile industry, truckers and technology workers. Depending

---

[15]   See summary of this case, and potential applications to gig economy issues, by Gibbins (2018).

[16]   The platform's unilateral control over the dispatch algorithm, and the opacity of how the algorithm assigns rides to respective drivers, certainly opens the possibility of this kind of retribution.

on occupation and jurisdiction, the mere owning of tools does not immediately disqualify workers from the normal protections afforded to other workers. For example, home workers in the textile industry, hairdressers in a salon and owner-operators in transportation or resource occupations are generally still entitled to be paid minimum wages and provided other normal entitlements of employment. Anyway, in a modern industrial economy, a passenger vehicle is an economically trivial productive asset. The value produced by ride-sourcing overwhelmingly reflects the drivers' labour, not their ownership of this small bit of capital. Annual depreciation on a typical vehicle (car) used for ride-source work (at most worth a few thousand dollars per year) represents a small share of the total value added in the course of a year's ride-source production. For comparison, the total net fixed private non-residential capital stock in the US economy (after depreciation) was estimated at over $25 trillion as at the end of 2018, or an equivalent of $165 000 per employed American.[17] The production process of ride-source services and other gig jobs is thus very labour-intensive relative to other jobs in the economy; and while the capital invested in drivers' vehicles may be significant to them personally, it is not significant in macroeconomic terms. Drivers cannot therefore be considered investors or owners in any meaningful sense. Indeed, there is a new class of digital businesses (such as HyreCar and SIXT) which rent vehicles to ride-source drivers who cannot afford their own car: it seems that those drivers have literally nothing to offer to the production process but their labour.

In labour law there is a long and complex jurisprudence as regards who constitutes an employee on strict legal grounds. (Chapter 4 in this volume provides a comprehensive discussion of employee tests, their history and evolution, and their implications for understanding and regulating gig work.) Typically, control is a critical dimension for determining a worker's status as an employee. For example, an important development in this area was a California law (Assembly Bill 5, or AB5; see State of California Department of Industrial Relations 2020) specifying an explicit three-part ABC test to determine employee status, including:

- Is the worker free from the control and direction of the hiring entity?
- Does the worker perform work outside the usual course of the hiring entity's business?
- Is the worker customarily engaged in an occupation or business of the same nature as that involved in the work performed?

---

[17] Author's calculations from US Bureau of Economic Analysis (2021, table 2.1) and US Bureau of Labour Statistics (2021).

Any work relationship which does not satisfy all of those tests in the affirmative would be defined as employment by default. It seems obvious that the level of direction and control exerted by Uber, Lyft and similar platform businesses qualifies their drivers as employees by this definition, and hence those businesses fought to defeat the California legislation.[18] The future evolution of jurisprudence regarding the employee test, and the vigour with which regulators apply it to digital platform businesses, is hard to predict. It will depend on the complex economic and political forces which the respective sides bring to bear in the continuing struggle over whether gig work deserves the same protections as other forms of wage labour.[19]

In an economic (if not legal) sense, however, it is hard to deny that workers for platforms such as Uber are employees. An Uber driver is told where to pick up passengers, and where to deliver them. They cannot set their fares, and they have no control over revenue (literally not even touching it). Uber can unilaterally change the revenue-sharing formula and other terms of engagement, control how fares are allocated to drivers and dismiss workers at will. Drivers must meet Uber's precise requirements regarding equipment, branding and service quality. The claim that these drivers are independent business operators on their own, that they are contracted by their paying customers and that Uber merely helps them operate their businesses, is not credible. That workers are not paid a fixed hourly wage, but only compensated a residual left from Uber-determined revenues after deducting all expenses (including Uber's unilaterally set fees) hardly proves these drivers are not workers; it implies, instead, that they are particularly vulnerable and exploited workers. It is not surprising, then, that this carefully constructed façade is being steadily pulled back by regulators, courts and policy-makers in jurisdictions around the world.

BACK TO THE ROOTS: HISTORICAL ANTECEDENTS FOR GIG WORK

We have identified four core common features of digital platform work. Platform companies:

• rely on contingent or on-demand labour;

---

[18] In 2020 the companies were ordered to transition their California drivers to employment relationships, but at the time of writing the companies were resisting the order, and had launched a plebiscite campaign to win exemption from the new law.

[19] Uber and other ride-source companies mobilised enormous financial resources, close to $200 million (Byrne 2020), to defeat the California law in a November 2020 plebiscite. This attests to their substantial economic interest in avoiding normal employer responsibilities.

- compensate labour through a piece-work system;
- require workers to provide their own capital equipment; and
- position themselves as an intermediary between producers and end-users.

The claim that these practices are recent innovations, facilitated by advances in technology, is historically false. None of these features is new; each has been applied in many industries, throughout the history of capitalism and wage labour (similar points are made by Finkin 2016 and Valenduc and Vendramin 2016).

For example, piece-work compensation is a long-standing practice in many industries, given its utility (in certain situations) for boosting productivity and ensuring that employers only pay for output they actually receive (Grantham 1994). So, the practice is not new, and it does not mean that employers can evade normal responsibilities (such as paying at least a minimum wage). Indeed, in most jurisdictions, piece-work compensation systems must ensure that workers earn at least the equivalent of the minimum hourly wage.[20] Also, while it is convenient for eliciting labour effort and discipline in some settings, piece work is not effective in most jobs. For maximum impact, piece-work compensation should be calculated at the individual level; it is unwieldy in jobs which require cooperation among teams. Moreover, in most jobs the output of labour is difficult to measure – more complicated than counting the number of widgets produced or the number of fast-food meals delivered. This measurement problem is especially acute where quality of output is important, not just quantity. So, piece work is neither a novel practice, nor a practice with potential to truly transform compensation in most existing jobs.

Similarly, requiring producers to supply their own tools and equipment has also been a common practice in many industries, even for waged employees. Most waged construction workers must purchase their own tools; waged hairdressers must supply their own equipment and hair products; and waged forestry workers must provide and maintain their own chainsaws and other personal equipment. Employers have long attempted to shift the cost and responsibility for individual tools and capital equipment to their workers. This effort, again, is neither novel, nor does it somehow transform the fundamental relationship between employers and their workers.

Contingent or on-demand work arrangements also have a long pedigree. It has been a common practice for many industries to staff labour on an

---

[20] In the USA, for example, compensation under piece rate systems must exceed the minimum wage calculated on an hourly basis (National Federation of Independent Business 2016). In the UK, payment per unit of output must exceed the minimum wage on an actual hourly basis, or on the basis of a fair rate of output (Government of the United Kingdom 2020).

on-demand basis, fluctuating with the flow of business. In earlier epochs the contingent workforce would gather each day at the workplace (for example, a mine, a wharf or a farm), hoping for an opportunity to work. In modern times, on-demand staffing could be facilitated through hiring halls,[21] labour hire agencies or other distanced technologies. Now it is facilitated through a digital app: faster and more efficient, but not structurally different.

The location of the digital platform business as an intermediary between the producer and end-user also has many historical precedents. Indeed, employers have long preferred (where possible) to constitute workers as contractors or nominally independent producers instead of as employees, for various economic and legal reasons: avoiding entitlements or benefits normally paid to employees, evading regulatory requirements regarding employment (such as minimum wages or limits on hours of work) and transferring risk of fluctuations in demand conditions to producers. Various forms of labour contracting were the predominant method for organising paid work in many industries in earlier periods of capitalism, including agriculture, resources and, even, heavy industry (Deakin 2000; Steinfeld 2001; Zmolek 2013). For example, the gangmaster system was a common form of wage labour in early capitalism, pioneered in agricultural work, but then extending into other sectors; it featured the mobilisation of short-term labour (often conscripted or indentured) to perform day-labour tasks, with a substantial proportion of the resulting compensation siphoned off by the broker or gangmaster. This system is still commonly used, even in many industrial countries (see Brass 2009; Melossi 2018), and some countries (including the UK and Australia) have enacted reforms to require the licensing of labour hire or gangmaster businesses.

All of these core features of modern gig work can be seen in long-standing employment practices from earlier decades and, even, centuries. Indeed, some earlier industries applied all of the core strategies now common in modern gig work. Consider, for example, the putting-out system of small-scale manufacturing in early European capitalism (also known as domestic or cottage production). The system was common in small-scale manufacturing sectors (such as textiles, clothing, footwear, cutlery, small furnishings and other

---

[21] After all, the term 'gig' originated in the performing arts industry, which recruited performers and support workers for short-term positions associated with specific productions. Performer, technician and stage-hand unions tried to regularise the process and support wages and conditions through union contracts and hiring halls. This industry thus provides both an early (pre-digital) example of gig work, and an early example of how workers can organise to improve wages and conditions within a gig context. See, for example, Williamson and Cloonan (2016).

simple consumer goods).[22] Merchants distributed production to dispersed workers, supplying them with necessary raw materials. Workers owned looms, lathes and other simple capital equipment, and worked from their own homes. The materials they worked on (including the ultimate finished products) were owned by the merchant capitalist, who hired them, organised production and marketed the output. Workers were compensated for their efforts on a piece-work basis, with payment occurring when the finished products were returned to the merchant. Except for the absence of digital methods for coordinating, supervising and compensating this work, the putting-out business model seems fully comparable to modern digital platforms.

In summary, apart from the application of digital methods of communication, work allocation, supervision and payment, the practices and relationships embodied in modern digital platform businesses do not seem new at all.

THE BROADER POLITICAL-ECONOMIC CONTEXT FOR GIG WORK: LESSONS FROM THE PAST AND IMPLICATIONS FOR THE FUTURE OF WORK

The central labour relationships embedded within modern digital platforms can thus be better understood as a return to previous practices, not as something truly original. However, while these contingent, on-demand staffing practices are not new, their importance has ebbed and flowed during history (Fudge 2017; Stanford 2017). For most of the twentieth century, employment relationships in many industries became more stable and, from the workers' point of view, secure. With the rise of Fordist mass production systems (in manufacturing and other sectors), technological and economic pressures pushed employers to organise a more reliable and disciplined workforce. After all, a large manufacturing facility, featuring an intense division of labour, requires a trustworthy workforce to be at their workstations at precise times, or else production cannot occur. The specialised skills associated with many jobs also required more predictable staffing strategies: employers had to recruit and train workers with reliable skills, and that is not feasible through contingent or on-demand models. Following the Second World War, there were additional macroeconomic and political motivations for the emergence of a more regularised and stable employment system. Unemployment was very low, and therefore recruiting contingent labour on an on-demand basis became harder. Employers came to value a stable workforce, even if that meant absorbing

---

[22]   For more detail on the putting-out or cottage systems, see Kriedte et al. (1981), Simonton (1998) and Mantoux (1961). The putting-out system still exists in some industries, such as watch-making.

some costs and risks associated with fluctuations in business demand. Strong workers' movements in many industrial countries after the war fought for improvements in labour standards, regulatory protections and social benefits; this also constrained employers' reliance on insecure labour strategies.

For a time, a more stable standard employment relationship (SER) emerged as a dominant employment norm.[23] More workers were employed in full-time, permanent positions, with relatively strong rights and entitlements. The development of a more extensive welfare system paralleled the rise of the SER, and many social benefits came to be attached to the status of recipients as waged workers. The SER was never universal: the system was highly gendered (women were often pushed into unpaid domestic work and/or less secure jobs in services industries) and racialised – immigrant workers never shared fully in the prosperity of the post-war era (Vosko et al. 2009). Nevertheless, the ability of employers to mobilise a contingent, on-demand workforce was constrained during this period, and for most workers jobs became more stable and predictable.

Under neoliberal economic and political governance, however, insecure and precarious work strategies have made a historic comeback. This has less to do with the technology of smartphones and digital platforms (as we have seen, contingent staffing strategies can be implemented in many low-technology settings), and more to do with the shifting balance of political-economic power. The unwinding of the SER is visible in many ways, not just via digital platform businesses. Indeed, evidence suggests that as little as half of existing paid work in some developed economies (such as the USA, Australia and Canada) still occurs within the confines of the SER.[24] Macroeconomic policy under neoliberalism deliberately re-created a permanent pool of unemployment, in part to undermine workers' wage demands and reinforce discipline in workplaces (Glyn 2006; Taylor and Ömer 2020). This facilitates contingent labour and gig work, which depend on a ready pool of desperate workers ready to log on at any time, for low and uncertain compensation. The stance of labour regulation also became more passive: where policy once actively promoted minimum labour standards and trade union activity, under neoliberalism it aimed to curtail union power, facilitate employer cost-cutting, and reinforce the compulsion to work by rolling back social benefits and income security.

These broad shifts in overall economic and political conditions, which have empowered employers to evade the protections and countervailing institutions

---

[23]   Vosko et al. (2003) and Bosch (2004) usefully catalogue the defining features of the SER.

[24]   See Gottfried (2014), Lewchuk et al. (2013), Independent Inquiry into Insecure Work (2012) and Carney and Stanford (2018) for more on the general expansion of precarious employment.

that emerged in the initial post-war decades, are vital to understanding the general growth in precarious work, and the rise of gig work in particular. It is not just technology which has facilitated the emergence of this new, particularly precarious form of employment. The whole direction of macroeconomic governance and labour regulation under neoliberalism has also been essential to the establishment and expansion of this model of employment.

CONCLUSION: THE UNCERTAIN FUTURE OF GIG WORK

Digital platform businesses utilise contingent staffing strategies that are hundreds of years old. These strategies have become popular among employers once again, not just thanks to digital technology, but also to compatible changes in economic, political and regulatory conditions. Gig work is expanding into new industries and occupations; recent data suggest it may soon account for one-tenth or more of total employment in industrial economies. Yet this business model faces constraints on its future expansion and viability, not just opportunities. As we conclude, let us consider several of the factors that will determine the future trajectory of gig work. These factors also indicate promising priorities for further research in this field.

Technology, Capital and Skill

Most jobs in a modern industrial economy require more sophisticated capital and tools than a car or a bicycle, and most jobs demand more specialised skills and qualifications than could be mobilised from a pool of anonymous gig workers. Digital platforms depend on workers who can do their jobs with no training, little capital equipment and no face-to-face supervision. Will low-skill, labour-intensive jobs such as these constitute a growing or shrinking share of future work? That depends on many economic and technological determinants. Limits on the proportion of work which can be performed in labour-intensive, lower-skill undertakings may constrain the future growth of gig work.

Piece-work Compensation

Paying gig workers on a piece-work basis is central to the business model of digital platforms, but there are limits to the usefulness of this form of compensation that may also curtail the future expansion of gig work. Piece-work compensation is most applicable to jobs performed on an individual basis (not by teams), and where quantity (not quality) of output is the primary metric of performance. Thus, while piece-work has been an important strategy for many employers through the history of capitalism, it has never become a universal system, for concrete economic and operational reasons. Today its applicability is still limited to a subset of jobs: work performed by individuals, that can be easily measured and in which quality factors are relatively unimportant. The limits to piece-work compensation may therefore also constrain the spread of gig work into other sectors and occupations.

Macroeconomic Conditions

The overarching state of labour market and macroeconomic conditions also shapes the success of contingent staffing strategies. When labour markets were very tight in the initial post-war decades, employers could not reliably recruit labour on a short-term basis; so they began to put more emphasis on ensuring workforce stability through permanent jobs and standardised employment relationships. The maintenance of full employment thus helped to underpin the expansion of the SER during that period. Under neoliberalism, however, with macroeconomic policy having abandoned full employment as a central goal (instead emphasising the maintenance of a permanent pool of unemployment to restrain inflation and discipline labour), conditions are more amenable to contingent staffing models. Widespread unemployment and underemployment facilitate on-demand and contingent staffing strategies in several ways: by suppressing wage growth, undermining worker expectations and ensuring a ready pool of labour available to meet the fluctuating demands of platform work. It seems likely that labour markets in industrial countries will take years to recover from the shock of the COVID-19 pandemic and associated recession. The consequent slack and desperation in the labour market will be highly conducive to the continuing expansion of contingent labour strategies of all kinds, including gig work.

Political and Regulatory Pressure

Platform businesses are facing serious regulatory and political challenges to their legitimacy, and their legal right to operate, in many jurisdictions. If these companies are ultimately forced to treat their workers as genuine employees, with normal rights and protections, they will not be viable in their current form.[25] Also, platform companies are aggressively resisting regulations that

---

[25]   Specifically, the unlimited ability of workers to sign on to work cannot be maintained if they must be guaranteed a minimum wage regardless of how much demand there is for their services. Treating workers as employees will require these businesses to begin to actively manage labour supply, limiting work to a number of workers commensurate with expected demand at various times.

would require them to provide workers with normal protections and entitlements. The outcome of these struggles will depend on the organisational and political power of the competing sides, and will shape the future expansion of on-demand platforms.

Financialisation

The growth and spread of digital platforms have also been facilitated by the extreme practices of financialisation. Hyperactive stock markets, shadow banking and other modern financial mechanisms have allowed companies such as Uber, which have yet to earn any profit, to nevertheless pay billions in capital gains to their founders and early investors. (Chapter 6 in this volume considers the importance of these financial mechanisms in further detail.) How do investors capture large gains from companies that are not profitable? Continuing inflows of finance (from other investors, from the commercial banking system or, even, provided by central banks through quantitative easing strategies) are required to sustain the capital market valuations of digital platforms despite their accumulating losses. How long that can continue is highly uncertain. Wider operating losses resulting from the COVID-19 pandemic could push some platform businesses into bankruptcy, undermining the financial inflows that have been so important to their growth. Alternatively, continuing injections of free money created by central banks might sustain the Ponzi-like momentum of high-technology equity markets for some time to come.

We have argued that the key labour practices utilised by digital platforms – on-call work, piece-work compensation, requiring workers to supply their own equipment, and an intermediated relationship between producers and end-users – are not new. Instead, these practices reflect a return to previous business strategies that were common in earlier periods of capitalism. Technology is not the only force behind this resurgence of precarious and contingent staffing strategies. Gig work has also been facilitated by a combination of economic and political circumstances, including the laxity of labour regulations, the persistence of unemployment and underemployment, and the reduced expectations of workers (especially young workers) that they can demand more from working life than an endless series of gigs. All of those factors can change, and therefore there is nothing inevitable (or technologically determined) about this mode of work. With more ambitious efforts to regulate gig work practices, with stronger commitments to full employment and income security, and with renewed efforts by working people to organise and campaign for decent jobs, the recent ascendancy of digital platforms and gig work could well be reversed in the future.

REFERENCES

Board of Governors of the Federal Reserve System (2019), *Report on the Economic Well-being of US Households in 2018*, New York: Board of Governors of the Federal Reserve System.

Bosch, G. (2004), 'Towards a new standard employment relationship in western Europe?', *British Journal of Industrial Relations*, **42** (4), 617–36.

Bowles, S. and H. Gintis (2008), 'Power', in L.E. Blume and S.N. Durlauf (eds), *New Palgrave Encyclopedia of Economics*, Basingstoke: Palgrave Macmillan, pp. 5077–82.

Brass, T. (2009), '"Medieval working practices"? British agriculture and the return of the gangmaster', *Journal of Peasant Studies*, **31** (2), 313–40.

Byrne, R. (2020), 'With funding from Uber, Lyft and Doordash, campaign behind California Proposition 22 tops $180 million', *The Center Square*, 9 September, accessed 7 April 2021 at https://www.thecentersquare.com/california/with-funding -from-uber-lyft-and-doordash-campaign-behind-california-proposition-22-tops-180 -million/article_d050ef74-f2c8-11ea-a5aa-d7e4a54c79a2.html.

Carney, T. and J. Stanford (2018), *The Dimensions of Insecure Work: A Factbook*, Canberra: Centre for Future Work, accessed 7 April 2021 at https://www.futurework .org.au/the_dimensions_of_insecure_work.

Deakin, S. (2000), 'Legal origins of wage labour: the evolution of the contract of employment from industrialisation to the welfare state', in L. Clarke, P. de Gijsel and J. Janssen (eds), *The Dynamics of Wage Relations in the New Europe*, Dordrecht: Springer, pp. 32–44.

Dow, G.K. (1993), 'Why capital hires labor: a bargaining perspective', *American Economic Review*, **83** (1), 118–34.

Farrell, D. and F. Greig (2016), *Paychecks, Paydays, and the Online Platform Economy: Big Data on Income Volatility*, New York: JP Morgan Chase Institute.

Finkin, M. (2016), 'Beclouded work in historical perspective', *Comparative Labor Law and Policy Journal*, **37** (3), 603–18.

Fudge, J. (2017), 'The future of the standard employment relationship: labour law, new institutional economics and old power resource theory', *Journal of Industrial Relations*, **59** (3), 374–92.

Gibbins, P. (2018), 'Extending employee protections to gig-economy workers through the entrepreneurial opportunity test of Fedex home delivery', *Washington University Journal of Law and Policy*, **57** (2), 183–204.

Glyn, A. (2006), *Capitalism Unleashed: Finance, Globalization, and Welfare*, Oxford: Oxford University Press.

Gottfried, H. (2014), 'Insecure employment: diversity and change', in A. Wilkinson, G. Wood and R. Deeg (eds), *The Oxford Handbook of Employment Relations*, Oxford: Oxford University Press, pp. 541–70.

Government of the United Kingdom (2020), 'Minimum wage for different types of work', accessed 7 April 2021 at https://www.gov.uk/minimum-wage-different-types -work/paid-per-task-or-piece-of-work-done.

Grantham, G. (1994), 'Economic history and the history of labour markets', in G. Grantham and M. MacKinnon (eds), *Labour Market Evolution: The Economic History of Market Integration, Wage Flexibility and the Employment Relation*, London: Routledge, pp. 1–26.

Henderson, T., T. Swann and J. Stanford (2018), *Under the Employer's Eye: Electronic Monitoring and Surveillance in Australian Workplaces*, Canberra: Centre for Future Work.

Independent Inquiry into Insecure Work (2012), *Lives on Hold: Unlocking the Potential of Australia's Workforce*, Melbourne: Australian Council of Trade Unions.

Jacobs, K. and M. Reich (2020), 'The effects of Proposition 22 on driver earnings: response to a Lyft-funded report by Dr. Christopher Thornberg', research brief, University of California Labor Centre, Berkeley, CA, accessed 7 April 2021 at https://laborcenter.berkeley.edu/the-effects-of-proposition-22-on-driver-earnings -response-to-a-lyft-funded-report-by-dr-christopher-thornberg/.

Jeon, S.-H., H. Liu and Y. Ostrovsky (2019), *Measuring the Gig Economy in Canada Using Administrative Data*, Ottawa: Statistics Canada, accessed 7 April 2021 at https://www150.statcan.gc.ca/n1/pub/11f0019m/11f0019m2019025-eng.htm.

Jost, M.P.S. (2011), 'Independent contractors, employees, and entrepreneurialism under the National Labor Relations Act: a worker-by-worker approach', *Washington and Lee Law Review*, **68** (1), 311–52.

Kosanovich, K. (2018), 'A look at contingent workers', Spotlight on Statistics, Bureau of Labor Statistics, US Department of Labor, Washington, DC, accessed 7 April 2021 at https://www.bls.gov/spotlight/2018/contingent-workers/pdf/contingent -workers.pdf.

Kriedte, P., H. Medick and J. Schlumbohm (1981), *Industrialization before Industrialization: Rural Industry in the Genesis of Capitalism*, Cambridge: Cambridge University Press.

Lewchuk, W., M. Lafleche, D. Dyson, L. Goldring, A. Meisner, S. Procyk, et al. (2013), 'It's more than poverty: employment precarity and household well-being', United Way, Toronto, accessed 2 April 2021 at http://www.unitedwaytyr.com/ document. doc?id=91.

Mantoux, P. (1961), *The Industrial Revolution in the Eighteenth Century*, New York: Harper and Row.

Manyika, J., S. Lund, J. Bughin, K. Robinson, J. Mischke and D. Mahajan (2016), 'Independent work: choice, necessity, and the gig economy', McKinsey Global Institute, San Francisco, CA, accessed 7 April 2021 at http://www.mckinsey.com/ global-themes/employment-and-growth/independent-work-choice-necessity-and -the-gig-economy.

Marshall, A. (1910), *Principles of Economics*, London: Macmillan.

McDonald, P., P. Williams, A. Stewart, R. Mayes and D. Oliver (2019), 'Digital platform work in Australia: prevalence, nature and impact', Queensland University of Technology, Brisbane, accessed 7 April 2021 at https://s3.ap-southeast-2.amazonaws .com/hdp.au.prod.app.vic-engage.files/7315/9254/1260/Digital_Platform_Work_in _Australia_-_Prevalence_Nature_and_Impact_-_November_2019.pdf.

Melossi, E.H. (2018), 'Labour contracting systems and African migrant networks in Foggia, Apulia, southern Italy', *Excursions*, **8** (1), 1–19.

National Federation of Independent Business (2016), 'What employers need to know about paying piece rate', accessed 7 April 2021 at https://www.nfib.com/content/ legal-compliance/legal/what-employers-need-to-know-about-paying-piece-rate -74647/.

Quinlan, M. (2012), 'The "pre-invention" of precarious employment: the changing world of work in context', *Economic and Labour Relations Review*, **23** (4), 3–24.

Robbins, L. (1932), *As Essay on the Nature and Significance of Economic Science*, London: Macmillan.

Samuelson, P. (1957), 'Wages and interest: a modern dissection of Marxian economics', *American Economic Review*, **47** (6), 884–921.

Simonite, T. (2015), 'When your boss is an Uber algorithm', *MIT Technology Review*, 1 December, accessed 7 April 2021 at https://www.technologyreview.com/2015/12/01/247388/when-your-boss-is-an-uber-algorithm/#:~:text=Whenper cent20aper cent20driverper cent20isper cent20logged,theirper cent20driverper cent20afterper cent20aper cent20ride.

Simonton, D. (1998), *A History of European Women's Work: 1700 to the Present*, London: Routledge.

Spencer, D. (2009), *The Political Economy of Work*, New York: Routledge.

Stanford, J. (2017), 'The resurgence of gig work: historical and theoretical perspectives', *Economic and Labour Relations Review*, **28** (3), 382–401.

Stanford, J. (2018), 'Subsidising billionaires: simulating the net incomes of UberX drivers in Australia', Centre for Future Work, Canberra, accessed 7 April 2021 at https://www.tai.org.au/sites/default/files/Subsidizing_Billionaires_Final.pdf.

State of California Department of Industrial Relations (2020), 'Independent contractor versus employee', Labor Commissioner's Office, Sacramento, CA, accessed 7 April 2021 at https://www.dir.ca.gov/dlse/faq_independentcontractor.htm.

Steinfeld, R.J. (2001), *Coercion, Contract and Free Labour in the Nineteenth Century*, Cambridge: Cambridge University Press.

Stewart, A. and J. Stanford (2017), 'Regulating work in the gig economy: what are the options?', *Economic and Labour Relations Review*, **28** (3), 420–37.

Taylor, L. and Ö. Ömer (2020), *Macroeconomic Inequality from Reagan to Trump*, Cambridge: Cambridge University Press.

Torpay, A. and A. Hogan (2016), 'Working in a gig economy', *Career Outlook*, US Bureau of Labor Statistics, Washington, DC, accessed 12 April 2021 at https://www.bls.gov/careeroutlook/2016/article/what-is-the-gig-economy.htm.

Uber Technologies Inc. (2019), 'Form S-1 registration statement', US Securities and Exchange Commission, accessed 7 April 2021 at https://www.sec.gov/Archives/edgar/data/1543151/000119312519103850/d647752ds1.htm#toc647752_11.

US Bureau of Economic Analysis (2021), 'Fixed assets accounts tables', National Income and Product Accounts, US Department of Commerce, Washington, DC, accessed 7 April 2021 at https://apps.bea.gov/iTable/iTable.cfm?ReqID=10&step=2.

US Bureau of Labor Statistics (2021), 'Labor force statistics from the current population survey, US Department of Labor, Washington, DC, accessed 7 April 2021 at https://www.bls.gov/cps/.

Valenduc, G. and P. Vendramin (2016), 'Work in the digital economy: sorting the old from the new', Working Paper 2016.03, European Trade Union Institute, Brussels, accessed 7 April 2021 at https://www.etui.org/publications/working-papers/work-in-the-digital-economy-sorting-the-old-from-the-new.

Vosko, L., N. Zukewich and C. Cranford (2003), 'Precarious jobs: a new typology of employment', *Perspectives on Labour and Income*, **4** (10), 16–26.

Vosko, L.F., M. MacDonald and I. Campbell (2009), 'Introduction: gender and the concept of precarious employment', in L.F. Vosko, M. MacDonald and I. Campbell (eds), *Gender and the Contours of Precarious Employment*, London: Routledge, pp. 1–25.

Williamson, J. and M. Cloonan (2016), *Players' Work Time: A History of the British Musicians' Union*, Manchester: Manchester University Press.

Zmolek, M.A. (2013), *Rethinking the Industrial Revolution: Five Centuries of Transition from Agrarian to Industrial Capitalism in England*, Leiden: Brill.

4. Labour protection for non-employees: how the gig economy revives old problems and challenges existing solutions

Victoria Daskalova, Shae McCrystal and Masako Wakui

REGULATING GIG WORK: A NOVEL PROBLEM?

In societies in which most people rely on their labour for their economic security, unregulated markets for labour can have significant consequences for working people, for the societies they live in and for the economy which depends their labour. An imbalance of power between individual workers and those who employ them can result in low wages and unsafe working conditions. Low wages, risk of injuries and the need to work, regardless of the circumstances, lead to precariousness – a term which has various definitions but roughly corresponds to feelings of uncertainty and insecurity about one's financial prospects and stability. Precariousness resulting from this imbalance of power is exacerbated when workers compete against each other. Cut-throat competition among workers can result in a race to the bottom, depressing wages and impacting working conditions. A society with high levels of poverty and precariousness suffers from many ills. A precarious workforce and a race to the bottom may also be harmful to industry itself in the long run, as over time it will be faced with a workforce which fails to improve its skills or which withers away or migrates.

Over the past two centuries, an elaborate legal framework has been developed in order to prevent or mitigate these problems and to regulate labour markets. This regulatory framework has been developed on the understanding that working people are at a structural and economic disadvantage in the markets for their labour when compared with those who hire them, and on recognition of the need for basic labour standards for the maintenance of acceptable living standards. This understanding has been entrenched within the international labour code, a set of core labour rights instruments maintained by the International Labour Organization (ILO), the only tripartite international institution responsible for the establishment of international norms and standards.

Gig work, which may be defined as the practice of parcelling work into short-term, discrete tasks that are outsourced to individuals under short-term contracts and remunerated on a per-task basis, has recently challenged the existing legal arrangements. This type of work has always been a feature of societies with well-regulated labour markets, but to a limited extent – think of the practice of individuals doing favours for each other in exchange for money or other benefits (for example, mowing your neighbour's lawn in exchange for payment or in exchange for baby-sitting). With the advent of digital platforms for labour or gigs, everything from jobs to everyday tasks can be parcelled out and sourced efficiently, whether from an extended local pool of workers or from a global pool of workers, as is the case of crowdwork and online gigs.

Gig work, at the time of writing, may be considered a relatively unregulated[1] market for labour. Questions such as, 'Are gig workers entitled to labour protection?', 'Can they bargain collectively?', 'Should they respect the rules governing enterprises or the rules governing workers?' are currently pending before courts and tribunals all over the world, and feature on the policy agendas of national and international policy-makers.

The issues gig work presents to policy-makers, industry and workers themselves resonate with the nineteenth-century problems that a constellation of norms, which today can be recognised as labour law, seek to regulate. This chapter provides an overview of these problems and aims to explain why the answers to societal problems, which seemed to work for so many years, have resurfaced in the context of the gig work phenomenon. In doing so, it shows what is new about regulating gig work in the twenty-first century and explores opportunities for addressing these regulatory challenges.

THE TRADITIONAL LEGAL SOLUTION: LABOUR LAW AND COLLECTIVE BARGAINING

Compared with the enterprises that engage them, workers are almost always in a weaker position in respect of access to capital (labour power cannot be stored but money and property can); subject to information asymmetry as regards knowledge of the market for their labour and the internal financial and operating status of the employing enterprise; and in respect of managerial decisions

---

[1]   Or, depending on the jurisdiction, under-regulated market for labour.

that could impact the security of their future employment (see Collins, 2000, pp. 11–16; Deakin and Wilkinson 2000).

The impacts of power imbalance between enterprises and workers occur across the life cycle of working relationships – at hiring, during engagement, and even after the relationship. For example, at the point of hiring, workers face significant information asymmetry between themselves and the enterprise. The enterprise has a better understanding of the market for the workers' labour, including the number of applicants for the position and their skill sets. The enterprise also understands its own financial position much better than does the applicant; the business may be struggling, and there may be risks of future redundancy or business failure that may make other work opportunities less risky. An individual worker will lack the resources and information necessary to counter the information held by the enterprise. Workers hired by the day in temporary or casual employment (work arrangements often described as precarious) have no guarantee of ongoing work and little ability to make long-term life decisions or investments. A worker in a more stable long-term work arrangement is also subject to power imbalance with the enterprise, owing to the degree of firm-specific investment the worker has made. Firm-specific skills may not easily be transferable, and if the worker has made significant life decisions on the basis of their steady income (house purchase, children, and so on), those obligations may not be sustainable if their employment comes to an end. Alternative work offers are difficult to obtain and, ultimately, labour markets are not competitive; if they were, worker mobility would be much greater (Steinbaum 2019, p. 6).

Most developed legal systems around the world now recognise the fundamental imbalance of power that exists between enterprises and their workers, but the forms of regulation that have developed in response differ in both complexity and content. However, they commonly involve some combination of minimum standards regulation whereby workers cannot be hired without being provided with certain terms and conditions, and implementation of the principles of freedom of association, enabling workers to act collectively to counterbalance the greater economic, managerial, information and property rights of employers (for useful comparative sources, see Blanpain 2014; Hendrickx n.d.; ILO 2012). The widespread inclusion of collective bargaining is recognised as the most appropriate regulatory technique for 'tackling the particular challenge of redistributing wealth and power between employer and employed' (Klare 2000, p. 84).

The implementation of collective bargaining as a solution to the imbalance of power between workers and employers has a long and complex history. In the UK, for example, there was early recognition by workers of the increased power they could exercise by acting collectively. However, the law took time to accept this possibility. Legislation passed by the UK Parliament in

1799 in the Combination Act made all worker combinations illegal. After the Combination legislation was eventually repealed in 1824, the UK courts found that worker collectives were inherently unlawful as a 'restraint of trade' (*Hornby* v. *Close* (1867) (30 Vict) 2 QB 153; *Farrer* v. *Close* (1869) (32 Vict) 4 QB 602; see further Elias and Ewing 1987, ch. 1; Wedderburn 1986, ch. 1). This undermined the very ability of workers to form collectives until the UK Parliament stepped in and passed the Trade Union Act 1871 (UK) which provided that collectives of workers were not inherently in restraint of trade, overcoming the judicial prohibition (for US history, see McCrystal 2007, pp. 15–17; Paul, 2016, p.1016). Further regulation in the UK over the twentieth century sought to level the playing field for collectives of workers to seek agreements with employers to improve their working conditions and achieve higher living standards (Wedderburn 1986, pp. 27–47). Similar histories arise across the developed world, of workers recognising the power of acting collectively and labour regulation gradually facilitating such action.

There has also been global recognition of labour rights through one of the oldest international organisations of the United Nations, the ILO.[2] From 1919, the ILO has pursued the implementation of a universally respected set of international labour standards, in the form of the international labour code, a code of workers' rights which has been developed in parallel to the protection of human rights through the modern UN system. The topics covered by the international labour code are broad and establish a complex set of interlocking minimum standards protection.[3] In particular, these standards require respect for the principles of freedom of association, which respect the right of all workers, without any distinction, to form organisations, and to act collectively in pursuit of their economic and social interests, including the right to engage in collective bargaining. The principles of freedom of association recognise the importance of self-determination and agency for working people, when setting the terms and conditions under which they labour.

Respect for the principles of freedom of association is a foundation principle of the ILO, and respect for the principles is embedded in its Constitution. Affirming this constitutional commitment, in 1998, the ILO Declaration on Fundamental Principles and Rights at Work confirmed that all member states of the ILO, even if they had not ratified relevant Conventions, have an obligation arising from being members of the ILO to respect, promote and realise, in good faith and in accordance with the Constitution, certain fundamental rights

---

[2]   The history of the ILO is set out in Alcock (1971) and current information about the organisation can be found at www.ilo.org (accessed 19 April 2021).

[3]   The standards are set out in the NORMLEX database maintained on the ILO website at www.ilo.org (accessed 15 April 2021).

(see further, Bellace 2001; Kellerson 1998). The first of those rights mentioned is freedom of association and the effective recognition of the right to collective bargaining.

While member states of the ILO are bound to respect the principles of freedom of association by virtue of their membership of the organisation, two Conventions further set out the substance of the principles themselves. Convention 87, the Freedom of Association and Protection of the Right to Organise Convention, expressly protects the right of workers and employers to establish and join organisations of their own choosing (article 2), and for those organisations to arrange their administration and activities and formulate their programs (article 3). Organisations for these purposes are 'any organisation of workers or of employers for furthering and defending the interests of workers or employers' (article 10). This principle of organisational autonomy set out in article 3 has consistently been found by the supervisory bodies of the ILO to extend to the right of worker organisations to strike in support of the economic and social interests of their members (ILO 2012, pp. 117–61, 2018, ch 10; see also, Creighton 2012; Novitz 2003, pt 4).[4]

Convention 98, the Right to Organise and Collective Bargaining Convention, protects union members against acts of anti-union discrimination (articles 1 and 2), meaning that they should not suffer any adverse consequences in respect of their employment or otherwise for having engaged in trade union activities or participated in collective bargaining or collective action. Article 4 commits member states bound by the Convention to encourage and promote the full development and utilisation of machinery for voluntary collective bargaining between employers or employers' organisations and workers' organisations, which is a positive obligation to establish and promote collective bargaining in domestic labour law systems (see further, Creighton 2012).

The principles of freedom of association protect the rights of workers and their organisations. The word 'worker' in this context applies broadly to all workers, not just to those who might be more narrowly defined as employees. The commentary of the ILO supervisory bodies endorse the proposition that the principles of freedom of association, and the relevant Conventions, extend to genuinely self-employed workers and those in disguised employment relationships (Creighton and McCrystal 2016 pp. 701–4). That is, the protections of the principles of freedom of association extend broadly to all working people, irrespective of whether they work as employees for a large enterprise at a factory, or if they are an entrepreneurial, self-employed worker, working only on their own account, and all those in between.

---

[4]   In recent years there has been some controversy at the ILO over this interpretation of Convention 87; see Bellace (2014, 2018); La Hovary (2013).

Despite the broad scope of international protections, the scope of protections within the international labour code are not routinely reproduced within domestic labour law systems. Domestic legal systems commonly provide association and bargaining rights to a narrower group of workers than those covered by the ILO principles of freedom of association. This boundary problem creates a siloing effect in practice between those workers who are in labour law's protective zone and those who are out and unable to act collectively in support of their economic and social interests. The boundary problem itself has always existed, but it has become more acute since changes to the manner in which work is organised in the late twentieth century and which is discussed in the next section (see, for example, Arup et al. 2006; Davidov and Langille 2006, 2011; Fudge et al. 2012; Stone and Arthurs 2013). These changes saw an increase in the number of workers engaged outside of the boundaries of labour law protection, with the most recent example of this being the rise of work in the gig economy.

PROBLEMS OF THE LEGAL SOLUTION

The rise of work in the gig economy, where workers are engaged through technology platforms that disconnect the labour engager from the hirer of that labour, challenge the capacity of existing labour law frameworks to adapt and provide adequate labour protection for the workers involved. The reasons for this are partly historical. They reflect the manner in which twentieth-century models of regulation are applied to forms of labour engagement that are reminiscent of daily labour-hire practices from the nineteenth and early twentieth centuries, mediated through modern technology; instead of lining up to wait for work at the start of the day, workers now log on to platforms to sit and wait.[5]

The systems of labour regulation that exist in the developed world are largely products of the mid-twentieth century – frameworks that were designed and implemented as part of the post-Second World War boom, and during a period of relative economic prosperity. As is well established within the literature, these models of regulation were developed at a time when enterprises were vertically integrated, with all necessary services employed in house. Workers were engaged in long-term relational contracts based on the assumption that workers would deploy their skills and experience for the benefit of the enterprise and under its control, and in return would receive opportunities

---

[5]   For the historical parallels between gig work and nineteenth-century forms of engagement, see Finkin (2016); Whiteside (2019).

for promotion and development, within the enterprise itself, and long-term employment security (see, for example, Collins 1990; Stone 2004, ch 2).

In response to underlying assumptions that this would continue to be the predominant model of engagement, labour law and social security systems were developed on the premise that the relationship of employment could be the locus of regulation. Labour rights would accrue to workers by virtue of these relationships, connecting to a workers' status as an employee, with some rights connected to the length of tenure of an employee in any particular enterprise (Stone 2004, ch 2).

However, the economic and structural models underpinning these labour and social security regulatory systems began to shift during the 1970s and 1980s. By the end of the twentieth century, many large vertically integrated firms had reorganised, with businesses devolving into more complex business networks and structures, where all but core functions were outsourced to other providers (see further, Johnstone et al. 2012, ch. 2; Stone 2004, pt 2; Weil 2017). These new business structures enabled the engagement of labour to be disconnected legally and practically from the enterprise that required the work to be undertaken (described in Johnstone et al. 2012, ch. 3). Complex supply chain networks developed to enable just-in-time manufacturing, with product parts produced by subcontractors and delivered only when required, thereby reducing the need for enterprises to carry the costs of large, permanent work-forces. The promise of long-term job security for workers was replaced with promises of skill development, which could be transferred from engagement to engagement, and workers were told to expect many different roles across their working lives (Collins 1990; Riley 2005; Weil 2017). The shift in business arrangements facilitated the rise of the supply of workers through labour hire arrangements (where workers may or may not be employed by the labour hire company) and in businesses seeking workers in self-employment to perform tasks, where enterprise risk and costs are shifted to the worker concerned. Work in the gig economy is the most recent variant on this fundamental shift in business operations where the disjunction between the company that needed work performed and the workers who would perform the work could be made even more clear by requiring those workers to bid for tasks through technology-based platforms (see Fudge 2006; Johnstone et al. 2012, ch. 3; Rosenblat 2018).

While enterprises have been restructuring and fundamentally changing the manner in which they engage labour, the laws regulating labour relationships have not kept up with these changes to ensure that the protections provided continue to apply to all workers. The rights and obligations that arise in many modern labour regulation systems remain connected to earlier assumptions over how labour is engaged. That is, those assumptions are based on the idea that workers receive benefits from engagement (permanency, entitlements, and so on) in return for surrendering control over how and when they perform work, and that this exchange justifies the application of protective labour regulation only for those who surrender control. Those who retain control must be in business for themselves. The gig economy challenges these underlying assumptions, constructing labour relationships that do not appear to fit the standard model of work engagement and which shift on to workers the risks of doing business, irrespective of the actual degree of dependency or control experienced by the worker.

At the core of this problem is the manner in which the legal systems in developed countries tend to create one category of protected worker who receives the full array of protective labour standards, and then leaves other workers to be regulated in the same manner as the enterprises that engage them, or creates lesser categories of workers with less protection than those in the protected class (see generally, Daskalova 2018; De Stefano 2016; Johnston and Land 2018). Those in the protected class of worker are most commonly labelled 'employees', and all rights and benefits flow from that label. The content of the definition of 'employee' then becomes the absolutely critical component in ensuring workers receive protection under labour rights and standards. Avoiding to engage workers as employees becomes the best way to avoid those more burdensome rights and standards. This approach has the effect in practice of making the legislative definition do all the work of protecting workers, even if the manner of its construction no longer matches the manner in which workers are engaged.

There is a great deal of variation in how the word 'employee' is defined in various legal systems around the world. However, at the core of the definition is commonly the central idea that employees labour for another, in the pursuit of the business objectives of another, and that this can be demonstrated in practice by a range of factors that show this, including, importantly, the degree to which the worker is subject to the control of that engaging enterprise (see generally, Casale 2011; Perulli 2003; Waas and Heerma van Voss 2017).

In Australia, for example, the difference between a contract of employment and other contracts under which work is performed, is the difference between a contract of service and a contract for services with the former, with the worker offering their service (not just their services) being employment. According to the High Court of Australia, the distinction being drawn between an employee and a contractor is 'rooted fundamentally in the difference between a person who serves his [sic] employer in his, the employer's, business, and a person who carries on a trade or business of his own' (*Hollis* v. *Vabu Pty Ltd* (2001) 207 CLR 21 at [40]). To determine this question, the courts apply what is known as a 'multi-factor' test (*Stevens* v. *Brodribb Sawmilling Co Pty Ltd* (1986) 160 CLR 16). This involves asking questions about the relationship between the labour engager and the worker to consider the aspects of the relationship that resemble employment, and those aspects that do not. The features that are considered are the right of the employer to exercise control (and to what degree), how the worker is paid, who provides the tools and equipment (ownership of assets), whether the worker can work elsewhere, or can turn down work with the engager, the ability to delegate work to others and the degree of integration into the business of the engager (see *Abdalla* v. *Viewdaze Pty Ltd* (2003) 122 IR 215, pp. 229–31). After considering the factors, a court must determine whether the factors support a conclusion that the worker is in law an employee, or is self-employed. In making this determination the degree to which certain factors influence the outcome in different cases may be highly fact specific (see further, Roles and Stewart 2012).

Approaches taken to determining the difference between an employee and other types of workers in other countries are similar, although with different degrees of emphasis on various factors. The UK courts for example, also consider the degree of control present in a working relationship as an important factor in assessing whether a relationship of employment exists, together with the degree to which a worker is integrated into the enterprise of the labour hirer and performs the work personally, and the reality of the work relationship when contrasted with the formal terms of any express contract (*Autoclenz* v. *Belcher* [2011] UKSC 41). The UK courts have also considered the degree of mutuality of obligation between the parties, that is, the extent to which the employer is required to provide work and the employee to perform it (Freedland and Dhorajiwala 2019, pp. 282–3).[6] In the USA, whether a worker is an employee is also a question of fact, by reference to a range of indicative factors including rights of control, the degree to which the worker is subject to direction and, potentially, the economic reality of the relationship and the extent to which the worker appears to be in business for themselves (Cherry 2016, p. 581–2). A further development is the ABC test being used by some US courts, which seeks to place the onus on the engager of a worker to establish that the worker is not working within their business operations but for themselves (see, for example, *Dynamex Operations West, Inc* v. *Superior Court of Los Angeles County*, Supreme Court of California, Ct.App. 2/7 B249546, 30 April 2018).

A similar approach applies in European countries within the civil law tradition. In France, for example, labour protections hinge upon a worker being engaged under a contract of employment, and the presence of a contract of employment is determined by reference to the facts of the case to determine if there is a legal relationship of subordination (Dockes 2019, pp. 220–21; Kessler

---

6   For discussion of the role of mutuality of obligation after *Autoclenz* v. *Belcher* [2011] UKSC 41, see McGaughey (2019). For the UK legal position more generally, see Jones and Prassl (2017, pp. 754–9).

2017, p. 204). This relationship is demonstrated in practice by showing that work is performed under the authority of an employer, shown through a range of factors including (but not limited to) managerial prerogative, the right to discipline a worker, the location of work duties (at the employer premises or not), the use of tools provided by an employer or the worker and exclusivity of engagement (Kessler 2017, pp. 204–5). In Italy, the Civil Code defines an employee as anyone who performs work in exchange for a wage in the view of cooperating within the company under the direction of the employer (Civil Code, Italy, art. 2094, cited in Ales 2017, p. 352; see also Gramano and Gaudio 2019, pp. 241–2). Whether this is satisfied in practice is a question of fact in each case (Ales 2017, p. 366). Out of this definition, the degree to which a worker is subject to the direction of the employer – subordination – is considered the most important component (Ales 2017, p. 364; Gramano and Gaudio 2019, pp. 241–2). Also, the law in France and Italy emphasises the importance of legal continuity in respect of the relationship between the hirer and the engager; that the worker is engaged and is expected to be present for work (Menegatti 2019; Gramano and Gaudio 2019, p. 242).[7]

This outline of the approaches taken in a range of jurisdictions reveals that there are common elements that arise in how the term 'employee' is approached. First, the question of whether a worker is an employee or not almost always depends on the facts of the particular case. There are generally no blanket rules that any type of worker is or is not an employee, and in each situation the court has to consider the facts of that case and weigh up the circumstances. Second, the question of whether a worker is an employee or not involves a group of factors, and how those factors are applied is different in different jurisdictions, so the weight given to different factors is not always the same. One factor that is very common is the extent to which a worker is subject to the control of the hirer in how they perform their work; that is, does the person hiring the worker have the authority to tell them how to do the work and the circumstances in which it will be done? The more control a hirer has over the work, the more likely it is that the worker is an employee. The other common factor is dependence; to what extent is the worker dependent on this particular hirer for work, or do they commonly work for other labour engagers? It is also significant whether there is an element of legal continuity in most jurisdictions; that is, an important component in the decision will be whether the worker is available to work when the labour engager says they must work (Menegatti, 2019).

---

[7] For the situation in Germany, see Schlachter (2019, pp. 230 33); for Spain see Todoli-Signes (2019, pp. 255–8).

Since assessment of whether a work relationship is employment or not is determined on the facts of each particular case, in practice the criteria lend themselves to manipulation through the ways enterprises who want work performed by workers enter into contracts with those workers for that work. Enterprises can organise their working relationships to minimise the factors that suggest control and dependence, and to emphasise elements of a working relationship that suggest the worker is independent or in business for themselves, even where this does not reflect the reality of the relationship. This process can be undertaken by shifting ownership and responsibility for maintenance of tools or assets to the worker, giving the worker a right to del-egate work or to work elsewhere, and minimising any appearance of control. However, the relationship as constructed on paper may not match the reality of how the worker is engaged and interacts with the labour engager; for example, a contractual right to delegate or work elsewhere may not be permitted in practice, or may not be economically possible.

When this legal position is considered in the context of the gig economy, we can see that it is arguable that gig economy platforms, with the willing or not-so-willing consent of workers, emphasise independence, flexibility and entrepreneurialism by workers. Workers are said by the platforms to be in business for themselves, with contact with their clients mediated through the platform. This manner of engagement picks up the factors in the employment tests most likely to lead to a finding that a workers is not an employee (see further, Cherry 2016; De Stefano 2016; Prassl 2018; Prassl and Risak 2016; Rosenblat 2018; Slee 2017; State of Victoria 2020).

If workers have no obligation to sign on to a platform at any particular time, that is, they have no obligation to be available for work, then this may defeat a finding of employment. Since platforms commonly depend on a large pool of unskilled labour, they can avoid requiring workers to log on to the platform at set times, and can incentivise workers to log on at peak times by increasing the rewards available, and potentially playing workers off against each other (Rosenblat 2018, ch. 2). If an employee is not required to be available to work on the command of the labour engager, then in many jurisdictions they may not be considered an employee. Further, a worker may find it necessary to sign on to several platforms at once to obtain enough tasks during the time they choose to work, which means that they appear to be running their own business, offering their services to a range of engagers. Transportation and food-delivery gig economy companies require workers to bear the costs of ownership and maintenance of assets, and running costs. Workers cannot bid for work on the platform if they do not have the right assets, and asset ownership is a tradi-tional sign of self-employment under definitions of employment. The practical realities of these relationships may involve a significant degree of control over the execution of work in practice (detailed guidance on how the work must be done, by whom and in what manner, combined with worker ranking systems and exclusion from the platforms for non-compliance). However, the construction of the relationships makes them appear as though the workers have far more independence than they are likely to have in reality, tending more easily towards findings that the workers are not employees under the traditional tests.

It is important to note that manipulating these tests to avoid a finding of employment is not new; labour engagers have been doing this in developed economies for decades. An example is cleaning staff who were one of the first group of workers to be outsourced (Berg et al. 2018, p. 29). Cleaners used to be employed by enterprises in house. By terminating the employment of in-house cleaners and putting cleaning work out to tender, enterprises broke the employment nexus, reconstructing workers as micro-businesses offering their services to multiple enterprises or diverting cleaning staff through labour-hire agencies (see Erickson et al. 2002). However, what is new in the gig economy is the use of an app to mediate the relationship between worker and labour engager, tapping into large groups of unskilled workers to perform unskilled labour, and dressing this up as entrepreneurialism on the part of the worker.

Where workers are not engaged as employees, they are excluded from labour regulation within a jurisdiction that relies on the definition of employee as the locus of labour-related rights. This can include, but is not limited to, minimum standards regulation (pay and minimum working conditions, such as annual, personal or parental leave), access to collective bargaining and strike action, access to workers' compensation protection, involvement in pension or superannuation plans, anti-discrimination protections and protection from arbitrary dismissal. In effect, workers are left to the same regulatory structures that apply to the enterprises that engage them – the laws of business, rather than the laws of labour. For workers seeking to combine to improve their working conditions, the largest obstacle presented by the laws of business is competition law and the cartel prohibition.

Misalignment with Other Areas of Law: The Problem of Competition Law and the Cartel Prohibition

By contrast to labour law regulation, where collective bargaining and the right to strike operate as a tool of worker empowerment, the underlying principles of modern competition policy and regulation perceive combination between enterprises as anti-competitive in nature. That is, competition policy dictates that in markets for goods and services, competition between enterprises will drive these enterprises to produce the goods and services in the most efficient manner, providing the goods and services to consumers at the most effective price point. If enterprises are able to act collectively, exchange information about their prices and methods of operation, or agree on, for example, price or territory distribution, this will interfere with the efficient operation of the market, impacting price signals and enabling enterprises to extract a higher price (a supra-competitive profit[8]) for their goods or services than they would otherwise be able to achieve in a competitive market.

However, the problem for workers in respect of competition laws arises when the promotion of freedom of association and collective bargaining for workers collides with the prohibition of cartels in competition law and regulation. The formation of trade unions and the exercise of collective bargaining rights for workers would otherwise constitute a cartel under competition law if the laws were applied to workers. However, for this reason, competition regulation generally contains a labour exemption, that is, carving out an exception for workers from the cartel conduct provisions and enabling workers' organisations to operate freely under labour laws without fear of prosecution under competition laws. However, as with all other labour protections, where the labour exemption within competition law is defined narrowly and covers only employee workers, those workers who fall outside the definition remain subject to competition regulation in the same manner as those who engage them. This means that if a gig economy worker is not covered by a labour exemption within the domestic competition regulation of a relevant jurisdiction, any collective action by that gig economy worker may be unlawful as a cartel under the relevant competition laws of the jurisdiction. While the laws of each jurisdiction are different, and the manner of application of cartel prohibitions are not uniform, the problem will remain; self-employed workers will be treated as enterprises for the purposes of competition regulation. This is a significant obstacle to the ability of those workers to act collectively and to seek to engage in collective bargaining.

BACK TO THE DRAWING BOARD? POLICY SOLUTIONS AND THEIR CHALLENGES

A number of regulatory solutions from a labour law perspective have been put forward to address these issues.

First, an approach would be for a reappraisal of how employment is defined in various legal systems so that it included workers in a broader range of circumstances and extended, potentially, to workers in the gig economy (see, for example, Stewart 2002). However, this would not be straightforward given the

---

[8]   Supra-competitive profit is profit beyond the profit which would occur in a competitive market.

extent to which different legal systems around the world have grappled with the difficulties of adequately defining an employee.[9]

A second approach is to create intermediate categories where gig economy workers are given some labour law protections, for instance, minimum standards or collective bargaining, but not others. For example, in the UK specific minimum labour standards are extended to those who fall under the definition of 'worker' in the Employment Rights Act 1996 (UK), and this was the basis upon which Uber drivers in the UK were found to be entitled to protection in respect of the national minimum wage and rights to paid leave (*Uber BV* v. *Aslam* [2019] IRLR 257; see Fredman and Du Toit 2019; Freedland and Dhorajuwala 2019, pp. 283–5; McGaughey, 2019).[10] Similarly, both Spain and Italy have systems to allocate some minimum standards rights to workers within defined intermediate categories (Spain: Landa Zapirain 2012; Sanchez Torres 2010; Italy: Ales 2017; Del Conte and Gramano 2018; Menegatti 2015). Another example is Canada where collective bargaining rights have been extended to dependent contractors, but not minimum standards rights or other labour protections (see Fudge et al. 2003; McCrystal 2014a). However, the problem with the intermediate categories approach is that giving a group of workers some rights, but not others, can create additional layers of complexity and confusion, potentially reproducing the types of definitional problems identified previously in respect of whether or not a worker is an employee (see Cherry and Alosi 2017; Stewart and McCrystal 2019).

A third approach is to develop sector-specific collective bargaining regimes for genuinely self-employed workers or for gig economy workers (see generally, McCrystal 2014b). This has been undertaken in Canada, where sector-specific models exist to facilitate collective bargaining by self-employed artists (Status of the Artist Act 1992 (Can); see Bernstein 2012), and for workers in the home day-care sector (Act Respecting the Representation of Certain Home Childcare Providers and the Negotiation Process for Their Group Agreements 2009 (Queb)). In Australia, a bargaining regime was implemented for truck owner-drivers working in the forestry sector in the State of Victoria (see Johnson 2012).

---

[9]   Lord Wedderburn famously referred to the tests for employment status as the 'elephant test', 'an animal too difficult to define but easy to recognize when you see it' (Wedderburn 1986, p. 116). However, it is worth noting that a recent inquiry by the Victorian State Government into the gig economy has made a series of recommendations for a statutory definition of employment that would cover gig workers that were not independent; see State of Victoria (2020).

[10]   Recent attempts to change the approach to the definition of employee and worker in the UK are outlined in Ewing et al. (2019).

The boundary problem is long-standing within labour law regulation, and these solutions have been well canvassed within the existing literature. A common feature of the solutions is their continued reliance on reframing the scope or coverage of labour law regulation in response to changes in work organisation. Given this focus, it is worth considering whether or not other solutions for regulating work in the gig economy might be found within the commercial realm, more specifically in the context of competition laws and policy.

Back to Collective Bargaining: Exemptions from Competition Law and their Limits

Given the potential difficulties of extending the employment concept to all gig workers and the problematic nature of creating intermediate categories of workers with discounted rights, collective bargaining may be a solution to enable gig workers to seek improved working conditions and exercise agency in their work relationships. Collective bargaining may achieve more flexible, industry-appropriate results than regulation will. However, owing to misalignment with competition law, in a number of jurisdictions collective bargaining has been legally constrained to those who fulfil the definition of employee or employee-like (sometimes termed 'worker'). A solution, therefore, might be to exclude collective bargaining by gig workers from the scope of competition law. However, this seemingly straightforward solution presents a number of normative and regulatory challenges from the perspective of competition regulation and policy. In order to understand the source of these difficulties and the challenges to finding a collective bargaining solution for gig workers, we must first understand the basic principles underpinning competition law and their evolution over time.

Competition law and labour law appear to have much in common. Both areas of law were primarily elaborated in the context of nineteenth-century developments; a time when advances in technology, business organisation, and developments in transportation and communication enabled the rise of large corporations. The industrial behemoths of two centuries ago exercised enormous economic and political power. In the USA, legal constructions known as trusts concentrated control over the main industries in only a few hands (Amato 1997). In this historical context, competition law, known as antitrust law in the USA due to its role in dissolving the previously mentioned trusts, was born. At its core, the goal of this new law was to ensure a level of de-concentration in the economy (cf. Bork 1978, who proposed that the goal was consumer welfare). Looking back to the origins of competition law, be it in nineteenth-century North America, or twentieth-century Europe, we see a concern with economic power and its ability to organise and entrench itself.

Already then, it became obvious that those in control of economic power were in a position to eliminate competitors, exploit their consumers and suppliers, and manipulate markets in order to optimise their private benefit to the detriment of the rest of the economy. The introduction of laws on competition was a reaction to this threat (Amato 1997; Hazlett 1992; Peritz 2000).

Considering this history, we may expect that competition law would play a complementary role to labour law. Labour law provisions, including collective bargaining provisions, aim to counterbalance the power of employers; antitrust law aims to prevent concentrations of power on markets. Competition law, thus construed, may be expected to act as an extra check on employer conspiracies and concentrations, thus improving the functioning of labour markets which suffer from a number of important imperfections (perishability, information asymmetries, lack of bargaining skills and inability to delay in negotiations). Rigorous enforcement of competition laws might therefore be expected to contribute to deconcentrated markets of employers, in which workers can choose among employers who compete on merit for their talent.

This logical expectation is not evidenced in reality. On the contrary, empirical evidence published over the past decade reveals that labour markets suffer from concentration and are not competitive in practice (Abel et al. 2019; Azar et al. 2017, 2018; Staiger et al. 2010). Scholars have also suggested that competition law enforcers have remarkably limited experience in addressing concentrations of employer power (Déchamps 2019; Marinescu and Hovenkamp 2019; Marinescu and Posner 2019; Naidu et al. 2018). Writing about the USA, Marinescu and Posner (2019) argue that antitrust law has failed workers by neglecting growing employer concentration leading to situations of monopsony[11] in markets for the purchase of labour. Recent antitrust investigations have revealed widespread and long-standing anti-competitive agreements among employers, such as agreements fixing employee wages or limiting the mobility of employees (US Department of Justice and US Fair Trade Commission 2016; US Department of the Treasury 2016; US White House 2016). This has raised the issue of why these practices have fallen under the radar of antitrust authorities for so long.

Adding to this meagre track record in protecting workers, antitrust law has come under criticism for its hostile approach to efforts by non-standard

---

[11] Monopsony is a model which represents a single purchaser sourcing an input from a market with many sellers. In this situation, the individual suppliers lack market power; the single purchaser, by contrast, has the power to set wages. More dynamic models of monopsony have been developed over time, which emphasise search costs, differentiation and other labour market frictions as the source of monopsony power. These markets may feature multiple purchasers which nonetheless have wage-setting power (Ashenfelter et al. 2010; Manning 2003).

workers, including gig workers, to organise and bargain collectively. The origins of this hostility may be a matter of ideology or ignorance (Naidu et al. 2018), but is also likely to stem from flaws in the legal framework (Masterman 2016). Over the course of the twentieth century, antitrust and labour law jurisprudence and legislation achieved a practical division of competences in order to avoid collective bargaining and the exercise of other worker rights (for example, boycotts and strikes) being caught by the antitrust provision. This division of competences often itself hinges on the concept of employee, which fails to cover the entire spectrum of self-employed workers.

The flaws of this solution have been highlighted by the rise of the gig economy. Notably, in the European Union (EU), competition law has been used by national competition authorities to block efforts by self-employed workers to organise (CEPS, Eftheia and HIVA-KU Leuven 2019; Daskalova 2018). For instance, in the Netherlands the competition authority has, until recently, taken the stance that self-employed workers do not benefit from an exemption for collective bargaining (Netherlands Competition Authority 2007, 2017; cf. Netherlands Competition Authority 2020) and, in Ireland, the competition authority has argued that self-employed voice-over actors would be in breach of competition law if they engage in collective negotiation with broadcasting companies (Irish Competition Authority 2004). Most recently, in Denmark, the competition authority has investigated the alleged anticompetitive effects flowing from a collective bargaining agreement involving platform workers. The investigation led to the platform withdrawing from the agreement (Danish Competition and Consumer Authority 2020). This leads to a paradoxical situation in which antitrust law, which is supposed to fight concentration, in practice is used to sanction arguably legitimate efforts to challenge concentration of power in the hands of labour engagers, which concentration antitrust itself has neglected. Unravelling this paradox requires tracing the way in which antitrust interpretation, implementation and enforcement have evolved over time.

Competition laws, although they may differ between jurisdictions, generally contain provisions on anti-competitive agreements, abusive conduct by powerful companies and merger control (control of concentrations). Collaboration between self-employed workers may be seen both as an example of an anti-competitive agreement, similar to a cartel, and in some instances even as an abuse of unilateral power. The economic impact of cartels is generally well established. A cartel limits output and, as a consequence, increases prices. In addition, since they eliminate competition among the parties, cartels are expected to lead to lower quality and less innovation than in a market with free competition. Cartels are considered so egregious that in many legal systems they are considered a *per se* infringement of competition, that is, conduct which is inherently harmful and therefore prohibited irrespective of the actual anti-competitive impact of the impugned behaviour.

Another relevant factor is the reorientation of the goals of competition law. Competition law doctrine has departed from a general concern with deconcentrated markets and fighting economic power, towards the more specific, but also narrower, objective of protecting efficiency and the welfare of consumers. These principles, ushered in with the Chicago revolution,[12] have gradually been adopted as a guiding principle of policy-making and enforcement in major jurisdictions such as the EU and the USA. The advent of this so named 'consumer welfare standard' has also influenced the priorities and methods of assessment employed by competition authorities. A consumer welfare standard specifically looks into the advantages (and especially the disadvantages) which accrue to consumers. The main problem with this approach is that it pits consumers against workers. Following the consumer welfare approach, an agency may conclude that collective bargaining outcomes, such as longer holidays, higher wages, and better safety and health protection at work, although possibly beneficial to society, are likely to be harmful to consumers because they raise costs of production and consumers end up paying more (Masterman 2016).

The assumption underpinning the consumer welfare standard is that higher wages and better benefits which increase worker welfare would ultimately have to be borne by consumers, thus diminishing consumer welfare. This crucial assumption does not hold in all cases: just as with higher costs for other inputs, to what extent higher labour costs are actually borne by the consumer depends on the economic conditions on the relevant downstream market and the position of the firm on this market (Masterman 2016). It is also worth stressing that the converse is not necessarily true; lower costs of labour achieved owing to the exercise of monopsony power do not necessarily translate into consumer benefit (Blair and Harrison 1991, 2010). That is, if a company saves on costs, it does not necessarily pass these savings on to its consumers, as it may be just as likely to return the profits to shareholders or investors. The likelihood that consumers will benefit also depends on the nature of the market and the

---

[12]   The Chicago revolution in antitrust refers to a change in approach to antitrust law and policy adopted by courts and agencies in the 1970s in the USA. In this period, which lasted until the 1990s, neoliberal economic ideas, espoused by members of the Chicago School of Economics influenced interpretation of competition law doctrine. It is generally considered that the influence caused a 'trimming back [of] antitrust doctrine' (Gerber 2010; Kovacic and Shapiro 2000, p. 54). In Europe, the Chicago School's influence was felt in the late 1990s and 2000s, when the European Commission introduced the modernisation package – soft law instruments emphasising consumer welfare as a value (Daskalova 2015).

position of the firm. Separating worker welfare from consumer welfare poses the question of difficult normative choices, which is why it has been suggested that worker welfare and consumer welfare should be seen as part of the same whole, with workers' welfare even being given a priority in specific cases (Masterman 2016). This argument is especially interesting in relation to gig workers engaged by platforms, as these workers may also be users of the platform.

Against this backdrop, it should be evident why exemptions from competition law are not so easily achieved. As a more general point, is worth re-evaluating the advantages and disadvantages of, respectively, a market with competition among workers and a market with cooperation between workers (collective bargaining) when considered through the lens of competition law and policy, which is focused on the efficient operation of markets. The following discussion outlines the arguments for and against restriction of the capacity to act collectively from that perspective in order to clarify the reluctance of some competition law academics, practitioners and authorities to entertain the possibility of an exemption.

When considering the competitive process, a helpful starting point is the realisation that competition occurs along a spectrum: there can be more competition or there can be less competition. On one end of the spectrum, there is no competition between workers but, at the other end of the spectrum, there is excessive competition, also known as a race to the bottom. Here, a distinction between product markets and labour markets is apposite. Severe competition among companies may result in company mergers or companies going bankrupt and exiting the market. However, whereas a company is a legal or financial concept, which can be founded and dissolved, a worker is a human being. Excessive competition among workers may result in everyone operating at or, even, below their cost (that is, the cost of human subsistence or the cost of dignified human subsistence). In this situation, only some workers earn a price premium (that is, marginal workers, who may, but need not, be the most efficient, excellent or poor workers).

Beyond implicit questions of fairness, a policy problem arises when workers provide labour below their subsistence minimum. A society in which workers live in poverty, unable to cover basic expenses, such as food, housing or shelter, and other necessities of a life consistent with human dignity (energy, water, transportation, education and access to culture), is an outcome which is arguably inconsistent with a society founded on respect for human rights. From a utilitarian perspective, the outcome is also unpalatable; if workers can literally not earn their living, they would have to be subsidised by the government or by private parties (for example, through charity, family networks or criminal activity). Thus, extreme competition among workers[13] is inconsistent with both the normative perspective informing legal reasoning and the utilitarian perspective espoused by economic sciences.

What possibilities are offered when competition among workers is replaced by cooperation? The other end of the spectrum, limited competition on price among workers owing to cooperation in the context of collective bargaining, is also worth evaluating. The normal concerns of competitive regulation in respect of the potentially anti-competitive effect of collective bargaining remain. In addition to the above-mentioned possible higher costs for employers (and perhaps also for end consumers), collective bargaining by workers may lead to some controversial results. Collective bargaining generally seeks to unify contract terms and conditions. If more value is placed on individual freedom than on solidarity, it may be considered undesirable to have unified terms and conditions. That is, a uniform wage may mean that some receive more than before, but it also may mean that some workers receive less than before (although in many systems collective outcomes operate as minimum standards). In order to do justice to these arguments, it is important to consider the welfare of all workers in the long run. Another controversial outcome is the situation in which genuine cooperation is compromised and when some members suppress other members' interests or promote solely their own benefit at the expense of those for others. In labour law regimes, express protection of the interests of minority groups is considered to be part of the regulatory design of bargaining models.

Another controversial approach towards unilateral worker initiatives is that which seeks to impose a price on employers or on final consumers. Consider, for instance, the widely publicised case of Uber drivers who agreed collectively to log out of the app simultaneously in order to induce surge pricing (Ingram 2017). Competition authorities, as well as consumers, may be less sympathetic to these unilateral forms of cooperation which are meant to advance the interests of one side (workers) and are not carried out transparently, cooperatively and in good faith. However, it is also possible that if these workers had already been provided with a workable collective bargaining option to seek to improve their hourly rates, they might not have seen the need to engage in conduct of this nature. That the workers concerned chose to act collectively to coordinate the collective logout means that those workers were *prima facie* open to the idea of acting in a collective manner.

---

[13] The implicit assumption here is that most persons earn their living from their labour. We might consider whether universal basic income, sufficient to cover the subsistence minimum, might solve the problem at least partially. Questions of fairness in dividing the gains from production between labour and capital remain.

Arguably different is the issue of how antitrust should appraise unilateral collective actions, such as strikes and boycotts. These collective initiatives may engender exclusionary effects, which hinder competition in the market, and may cause some disruption to the operation of the economy as well as society. Note that anti-competitive collective actions and, in particular, collective boycotts harm not only the employer's interests but also the workers' interests in relation to loss of income and, ultimately, loss of livelihood. Owing to its nature, a collective boycott, and thus the level of anti-competitive detriment, is constrained by both an internal factor (loss of source of revenue) and an external factor (alternative contractors for the employers). What approach competition law should take to this cooperation remains a matter of debate.

By contrast, it is also recognised that collective bargaining can produce public benefits or even have pro-competitive effects. Collective bargaining can be efficient in saving transaction costs, reducing information asymmetry between the parties and enabling the less powerful party to have greater input in the negotiation process. This is particularly so where a powerful buyer unilaterally sets the standard form of contracts and offers it to weaker counter-parties on a take-it-or-leave-it basis. Other issues that could be addressed collectively include the monitoring and enforcement of contracts. When an agreed contract is not honoured, the aggrieved party often remains silent owing to a lack of expertise and resources, as well as fear of retaliation. The situation is not only unfair but also inefficient as this situation may discourage workers from making the optimal investment in the employment relationship. Acting together can ease these problems in monitoring compliance; bringing legal suits becomes much easier when undertaken collectively. Collective bargaining undertaken by the group, composed of these workers or self-employed, and the procurer of their services or products, may be an effective way to protect weaker parties as well as to achieve potential efficiency (Dau-Schmidt and Traynor 2009; Kaufman 2004).

Collective bargaining may have advantages and disadvantages, but so does excessive competition among workers. Given the concentration of power on the employer side and the revelations about widespread anti-competitive practices by employers, we may view collective bargaining as a corrective mechanism, instead of as a restrictive mechanism. Nonetheless, difficulties remain. Competition law experts continue to struggle with fine-tuning the scope of an exemption and striking the right balance between competition and cooperation among workers. The concept of employee as a dividing factor arguably has reached its limits. This raises the question, what if it were possible to offer protection to platform workers, without trying to stretch the definition of 'employee', simply by extending the right to collective bargaining to them?

When thinking about how different competition agencies have approached the issue of exemptions and who should be able to engage in horizontal coordination, Australian competition law is an outlier in respect of the broad potential for collective bargaining to occur by groups of small business actors (including the self-employed). Since 1974, the relevant Australian legislation has allowed the regulator, the Australian Competition and Consumer Commission (ACCC), to authorise collective conduct by enterprises that are otherwise in competition with each other, where it can be shown that the conduct would have a net public benefit (Competition and Consumer Act 2010 (Cth) s 90(7)). This process has been augmented in practice since 2007 by a notification procedure which allows businesses proposing to engage in conduct with low transaction values to notify the ACCC of their intention to engage in collective bargaining, which conduct can then proceed free of legal liability unless the ACCC objects (Competition and Consumer Act 2010 (Cth) s 93AC).

The underlying assumption behind the Australian approach is that horizontal coordination between small business actors where they are seeking to engage as a collective with a counter-party, a target for collective bargaining (as opposed to simply agreeing prices between themselves), is unlikely to produce substantial anti-competitive effects. On the contrary, the contractual terms produced by these arrangements are likely to be of public benefit through efficiency gains where the contracting arrangements result from better informed bargaining parties who are better represented, armed with better and more information, and are all acting voluntarily.[14] In recognition of this fact, the ACCC is currently pursuing a 'collective bargaining exemption' whereby small businesses (with an aggregated turnover of less than $10 million) will be able to form collectives and seek to bargain collectively with a target, on the assumption that these arrangements generally produce public benefit (and therefore are not necessarily anti-competitive) (ACCC 2019; Hardy and McCrystal 2020).

The aspects of the ACCC approach that are significant are the manner in which it challenges the assumption that all horizontal coordination is necessarily anti-competitive (in particular where that coordination involves small actors bargaining with a counter-party), and it avoids the need to label the parties as 'employees', 'workers' or 'the self-employed'. Business turnover is a simpler measure than contractual status and does not rely on judicial interpretation.

The Australian example also supplies some evidence that providing exemptions for collective bargaining by small enterprises is not the beginning of a downward slippery slope. The move to a general exemption, after decades of

---

[14] For the definition of 'public benefit', see *Re 7-Eleven Stores Pty Ltd, Australian Association of Convenience Stores Incorporated and Queensland Newsagents Federation* [1994] ATPR [41-357], p. 17 242 and McCrystal (2009).

the authorisation model and 12 years of notifications, has been partly prompted by the low take-up rates under the collective bargaining notification provisions. Despite fears expressed before their introduction that the notification provisions would undermine the emphasis in competition regulation on preventing cartels (Dawson et al. 2003, p. 118), the take-up rate of notifications has been around five per year, which hardly represents any kind of deluge.[15]

A final aspect of the Australian provisions worth noting is that the ACCC has the power to authorise collective boycotts (that is, strikes), and small businesses have the capacity to notify the ACCC of proposed collective boycott conduct. However, the provisions have almost never been used. In fact, recent ACCC literature on collective bargaining suggests that in particular circumstances collective boycotts may promote efficiency by bringing parties to the bargaining table in order to allow collective bargaining to take place (ACCC 2018, pp. 10–11; see also Hardy and McCrystal 2020). This type of conduct has not been included within the proposed exemption.

The previous evidence shows how a simple notification procedure with appropriate thresholds and safeguards, such as voluntariness, good faith and prohibition of unilateral tariff-setting, can reconcile labour law and competition law objectives. Importantly, the system has not resulted in a flood of notifications, nor has it led to an erosion of competition law standards. It is curious to see whether other jurisdictions considering these exemptions, for example, the EU (European Commission 2020), will follow Australia's example.

Complementary Legislation: Data Protection and Platform Regulation

Where workers are not engaged as employees, they are excluded from labour regulation within a jurisdiction that relies on the definition of employee relating to collective bargaining and strike; they are also excluded from other laws regulating employment contract, which can include, but are not limited to, minimum standards regulation (pay and minimum working conditions, such as annual, personal or parental leave), access to workers compensation protection, involvement in pension or superannuation plans, anti-discrimination protections and protection from arbitrary dismissal. Competition law can play a more active role in scrutinising abuses of power, such as unfair terms (including excessively low wages, unfair refusal to deal, discriminatory wages and terms of work, and exclusionary provisions limiting the possibility to work for a competitor). However, competition law also has limitations. Complementary

---

[15]   All authorisations and notifications are published on a public register available at www.accc.gov.au (accessed 19 April 2021). This figure was reached by averaging the number of notifications over the period of the operation of the notification provisions.

pieces of regulation, for example, on fairness in platform-to-business rela-tionships or on user rights (as opposed to only worker rights), may create additional layers of protection for platform workers, thus helping fill gaps in the legal framework. This complementary legislation, such as Regulation 2016/679 in the EU, better known as the General Data Protection Regulation (GDPR), and Regulation 1150/2019 on promoting fairness and transparency for business users of online intermediation services, can facilitate enforcement of rights for platform workers. For instance, the GDPR enables individual plat-form workers to request their personal data that the platform has accumulated. The data can be helpful in clarifying related lawsuits regarding, for example, status as an employee or other rights. The lawsuit of two Uber drivers in the UK who demanded access to their personal records held by the company is illustrative of the role which can be played by complementary pieces of legis-lation (CPI 2019).

Platform-specific regulations can also play a complementary role, as long as they recognise the special position of workers in the platform economy. In the EU, Regulation 1150/2019 on promoting fairness and transparency for busi-ness users of online intermediation services requires basic procedural justice, such as: clearly stated terms of contract, especially regarding suspension and termination, and available services; notice periods in case changes are made to the contract and prohibition of retroactive changes to contracts; a notice period and statement of reasons in case a user's access to the platform ser-vices is restricted, suspended or terminated; an unambiguous statement of the parameters influencing ranking of business users and regarding access to data; and clarity, including justification, regarding possible differentiated treatment of different business users by the platform. The provisions of this regulation promise to raise the quality of contracts by ensuring that platform users have access to the information they need in order to make a rational assessment of the economic environment in which they operate. However, the platform regulation was introduced with corporate users in mind (for example, hotels providing rooms via booking websites, distributors selling products via online platforms and developers distributing apps via app stores); the effect of this regulation on platform workers is not yet known. Importantly, the regulation applies only to business users, that is, users who provide goods or services to final consumers (articles 1(2), 2(1) and 2(4) of Regulation 1150/2019). The regulation does not apply to platform users who provide services to other businesses; therefore, whole segments of platform workers, for example, those engaged in crowdwork projects for a business client and possibly even food deliverers who dispatch meals from local restaurants, fall outside the scope of this regulation.

Despite these promising developments, it must be recognised that, as things stand, there is a limit regarding the extent to which competition law, as currently interpreted and enforced, and complementary legislation can fulfil the functions of labour law regarding unfair and abusive practices. Competition law with its suspicion of cartels, focus on consumer welfare, requirements regarding evidence and rigorous methodology of economic assessment, is not suited to address the day-to-day concerns of a number of workers. Furthermore, the comparison is not one-to-one, as labour law has a more elaborate system covering matters from solicitation, to retirement, than competition law has. Given that labour law regulates a particular market, for human labour, it takes into account, much like sector-specific regulation, the specific needs of humans. General competition law, by contrast, does not delve into matters concerning, for example, maternity leave, disability or training at work. Nor does competition law address fundamental rights issues connected with human labour, such as the right to freedom of expression at work or data justice and privacy at work. With proper calibration and enforcement, competition law and similar legal tools, if appropriately enforced, may help complement labour law achieve its goals on markets for labour; however, currently, this is not the case.

CONCLUSION

Labour law, collective bargaining and social security systems are part of an intricate system aiming to ensure that workers avoid poverty and exploitation, and that they receive a fair share of their labour. A platform economy relying on workers in non-standard employment is upsetting the delicate balance between competition and solidarity achieved in this system. In welfare states, the employment relationship is the gateway to benefits, insurances and social protection. Collective bargaining provides extra influence as well as the collective wisdom of a community within which individual interests can be protected. A label which places workers in the realm of commercial law means the worker is subject to all the risks and uncertainties of economic life – fluctuations of demand, vigorous competition, insolvency and liability. Unlike a company which can go bankrupt, be resold or merged with another when its existence is no longer supported by the market, workers who are faced with the consequence of losing their job, still have to find a way to survive. The market does not correct itself.

As for the platform economy, some authors already write of super-competition among platform workers (Drahokoupil and Piasna 2017), whereas others have found evidence of monopsony power on online labour platforms (Dube et al. 2020). Some platform workers may be sceptical about being represented by traditional labour unions, but all workers resent falling rates of compensation. If platform concentration and worker competition exacerbate, demands for protection will likely increase, and the question of employment status and right to collective bargaining will stay relevant.

This chapter has not touched on other important considerations, such as fundamental rights at work (for example, free speech, gender equality and protection for persons with disabilities). The same can be said of issues related to safety and liability. Often it is the 'employee' label that determines whether (and to what extent) platforms have obligations regarding non-discrimination and equal treatment, harassment, safety and liability. The employment relationship is often the locus of extensive public and private regulation, some of it directly related to the performance of the job at hand, for example the need for safety precautions, and some of it related to broader societal concerns, such as the obligation to train employees and help them retain marketable conditions.

There is a great deal at stake with labels. If the platform economy is to expand in a meaningful way, important legal re-engineering has to take place to ensure that workers are treated fairly and are not forced to compete with each other in unsustainable ways.

REFERENCES

Abel, W., S. Tenreyro and G. Thwaites (2019), 'Monopsony in the UK', CEPR Policy Portal, accessed 19 April 2021 at https://voxeu.org/article/monopsony-uk.

Alcock, A. (1971), *History of the International Labour Organisation*, London: Macmillan.

Ales, E. (2017), 'The concept of "employee": the position in Italy', in B. Waas and G.H. van Voss, (eds), *Restatement of Labour Law in Europe: The Concept of the Employee*, Oxford: Hart, pp. 351–75.

Amato, G. (1997), *Antitrust and the Bounds of Power: The Dilemma of Liberal Democracy in the History of the Market*, Oxford: Hart.

Arup, C., P. Gahan, J. Howe, R. Johnstone, R. Mitchell and A. O'Donnell (eds) (2006), *Labour Law and Labour Market Regulation*, Sydney: Federation Press.

Ashenfelter, O.C., H. Farber and M.R. Ransom (2010), 'Modern models of monopsony in labor markets: a brief survey', IZA Discussion Paper No. 4915, Institute of Labor Economics, Bonn, pp. 1–12.

Australian Competition and Consumer Commission (ACCC) (2018), 'Small business collective bargaining: Notification and authorisation guidelines', December, ACCC, Canberra.

Australian Competition and Consumer Commission (ACCC) (2019), 'Class exemption for collective bargaining, guidance note', draft for consultation, June, ACCC, Canberra.

Azar, J., I. Marinescu and M.I. Steinbaum (2017), 'Labor market concentration', NBER Working Paper No. 24147, National Bureau of Economic Research, Cambridge, MA.

Azar, J., I. Marinescu, M.I. Steinbaum and B. Taska (2018), 'Concentration in US labor markets: evidence from online vacancy data', IZA Discussion Paper No. 11379, Institute of Labor Economics, Bonn, accessed 19 April 2021 at http://ftp.iza.org/dp11379.pdf.

Bellace, J. (2001), 'The ILO declaration of fundamental principles and rights at work', *International Journal of Comparative Labour Law and Industrial Relations*, **17** (3), 269–87.

Bellace, J. (2014), 'The ILO and the right to strike', *International Labour Review*, **153** (1), 29–70.

Bellace, J. (2018), ILO Convention No. 87 and the right to strike in an era of global trade', *Comparative Labor Law and Policy Journal*, **39** (3), 495–530.

Berg, J., M. Aleksynksa, V. De Stefano and M. Humblet (2018), 'Non-standard employment around the world: Regulatory answers to face its challenges', in F. Hendrickx and V. De Stefano (eds), *Game Changers in Labour Law: Shaping the Future of Work*, Alphen aan den Rijn: Kluwer Law International, pp. 27–48.

Bernstein, S. (2012), 'Sector-based collective bargaining regimes and gender segrega-tion: a case study of self-employed home childcare workers in Quebec', in J. Fudge, S. McCrystal and K. Sankaran (eds), *Challenging the Legal Boundaries of Work Regulation*, Oxford: Hart, pp. 213–29.

Blair, R.D. and J.L. Harrison (1991), 'Antitrust policy and monopsony', *Cornell Law Review*, **76** (2), 297–340.

Blair, R.D. and J.L. Harrison (2010), *Monopsony in Law and Economics*, New York: Cambridge University Press.

Blanpain, R. (ed.) (2014), *Comparative Labour Law and Industrial Relations in Industrialized Market Economies*, 11th edn, Alphen aan den Rijn: Kluwer Law International.

Bork, R.H. (1978), *The Antitrust Paradox: A Policy at War with Itself*, New York, Basic Books.

Casale, G. (ed.) (2011), *The Employment Relationship: A Comparative Overview*, Geneva: International Labour Office.

Centre for European Policy Studies, Eftheia and Haplarithm inference of Variant Alleles-KU Leuven (CEPS, Eftheia and HIVA-KU Leuven) (2019), *Study to Gather Evidence on the Working Conditions of Platform Workers, Final Report VT/2018/032*, report prepared for the European Commission, 10 November.

Cherry, M. (2016), 'Beyond misclassification: the digital transformation of work', *Comparative Labor Law and Policy Journal*, **37** (3), 577–602.

Cherry, M. and A. Alosi (2017), '"Dependent contractors" in the gig economy: a com-parative approach', *American University Law Review*, **66** (3), 635–89.

Collins, H. (1990), 'Independent contractors and the challenge of vertical disintegration to employment protection laws', *Oxford Journal of Legal Studies*, **10** (3), 353–80.

Collins, H. (2000), 'Justifications and techniques of legal regulation in employment', in H. Collins, P. Davies and R. Rideout (eds), *The Legal Regulation of the Employment Relation*, London: Kluwer Law International, pp. 3–27.

Competition Policy International (CPI) (2019), 'UK: drivers file GDPR lawsuit against Uber', 25 March, accessed 19 April 2021 at https://www.competitionpolicyint ernational.com/uk-drivers-file-gdpr-lawsuit-against-uber-2/.

Creighton, B. (2012), 'International labour standards and collective bargaining under the *Fair Work Act 2009*', in B. Creighton and A. Forsyth (eds), *Rediscovering Collective Bargaining*. New York: Routledge, pp. 46–67.

Creighton, B. and S. McCrystal (2016), 'Who is a worker in international law?', *Comparative Labor Law and Policy Journal*, **37** (3), 691–725.

Danish Competition and Consumer Authority (2020), 'Commitment decision on the use of a minimum hourly fee', accessed 19 April 2021 at https://www.en.kfst

.dk/nyheder/kfst/english/decisions/20200826-commitment-decision-on-the-use-of-a -minimum-hourly-fee-hilfr/.

Daskalova, V.I. (2015), 'Consumer welfare in EU competition law: what is it (not) about?', *The Competition Law Review*, **11** (1), 133–62.

Daskalova, V.I. (2018), 'Regulating the new self-employed in the uber-economy: what role for EU competition law', *German Law Journal*, **19** (3), 461–508.

Dau-Schmidt, K. and A. Traynor (2009), 'Regulating unions and collective bargaining', in K. Dau-Schmidt, S. Harris and O. Lobel (eds), *Labor and Employment Law and Economics*, Cheltenham, UK and Northampton, MA, USA: Edward Elgar, pp. 96–128.

Davidov, G. and B. Langille (eds) (2006), *Boundaries and Frontiers of Labour Law*, Oxford: Hart.

Davidov, G. and B. Langille (eds) (2011), *The Idea of Labour Law*, Oxford: Oxford: Oxford University Press.

Dawson, D., J. Segal and C. Rendall (2003), *Review of the Competition Provisions of the Trade Practices Act*, Canberra: Commonwealth of Australia.

De Stefano, V. (2016), 'The rise of the just-in-time workforce: on-demand work, crowdwork, and labor protection in the gig-economy', *Comparative Labor Law and Policy Journal*, **37** (3), 471–504.

Deakin, S. and F. Wilkinson (2000), 'Labour law and economic theory: a reappraisal', in H. Collins, P. Davies and R. Rideout (eds), *The Legal Regulation of the Employment Relation*, London: Kluwer Law International, pp. 29–62.

Déchamps, P. (2019), 'Labour markets: a blind spot for merger control?', *Oxera Agenda*, September, accessed 19 April 2021 at https://www.oxera.com/agenda/ labour-markets-a-blind-spot-for-merger-control/.

Del Conte, M. and E. Gramano (2018), 'Looking to the other side of the bench: the new legal status of independent contractors under the Italian legal system', *Comparative Labor Law and Policy Journal*, **39** (3), 579–606.

Dockes, E. (2019), 'New trade union strategies for new forms of employment', *European Labour Law Journal*, **10** (3), 219–28.

Drahokoupil, J. and A. Piasna (2017), 'Work in the platform economy: beyond lower transaction costs', *Intereconomics*, **52** (6), 335–40.

Dube, A., J. Jacobs, S. Naidu and S. Suri (2020), 'Monopsony in online labor markets', *American Economic Review: Insights*, **2** (1), 33–46.

Elias, P. and K. Ewing (1987), *Trade Union Democracy, Members' Rights and the Law*, London: Mansell.

Erickson, C.L., C.L. Fisk, R. Milkman, D.J.B. Mitchell and K. Wong (2002), 'Justice for janitors in Los Angeles: lessons from three rounds of negotiations', *British Journal of Industrial Relations*, **40** (3), 543–67.

European Commission (2020), 'Competition: the European Commission launches a process to address the issue of collective bargaining for the self-employed', Press Release IP/20/1237, 30 June, accessed at https://ec.europa.eu/commission/ presscorner/detail/en/IP_20_1237.

Ewing, K., J. Hendy and C. Jones (2019), 'The universality and effectiveness of labour law', *European Labour Law Journal*, **10** (3), 334–8.

Finkin, M. (2016), 'Beclouded work in historical perspective', *Comparative Labor Law and Policy Journal*, **37** (3), 603–18.

Fredman, S. and D. Du Toit (2019), 'One small step towards decent work: *Uber v Aslam* in the Court of Appeal', *Industrial Law Journal*, **48** (2), 260–77.

Freedland, M. and H. Dhorajiwala (2019), 'UK response to new trade union strategies for new forms of employment', *European Labour Law Journal*, **10** (3), 281–90.

Fudge, J. (2006), 'Fragmenting work and fragmenting organizations: the contract of employment and the scope of labour regulation', *Osgoode Hall Law Journal*, **44** (4), 609–48.

Fudge, J., S. McCrystal and K. Sankaran (eds) (2012), *Challenging the Legal Boundaries of Work Regulation*, Oxford: Hart.

Fudge, J., E. Tucker and L. Vosko (2003), 'Employee or independent contractor? Charting the legal significance of the distinction in Canada', *Canadian Labour and Employment Law Journal*, **10** (1), 193–230.

Gerber, D.J. (2010), *Global Competition: Law, Markets, and Globalization*, New York: Oxford University Press.

Gramano, E. and G. Gaudio (2019), '"New trade union strategies for new forms of employment": focus on Italy', *European Labour Law Journal*, **10** (3), 240–53.

Hardy, T. and S. McCrystal (2020), 'Bargaining in a vacuum? An examination of the ACCC's collective bargaining class exemption', *Sydney Law Review*, **42** (3), 311–42.

Hazlett, T.W. (1992), The legislative history of the Sherman Act re-examined. *Economic Inquiry*, **30** (2), 263–276.

Hendrickx, F. (ed.) (n.d.), *International Encyclopaedia for Labour Law and Industrial Relations*, Wolters Kluwer, Kluwer Law Online, accessed 19 April 2021 at https://kluwerlawonline.com/Encyclopedias/IEL+Labour+Law/740.

Ingram, K. (2017), 'Uber drivers are gaming the system and even going offline en masse to force "surge" pricing', *University of Warwick News and Events*, accessed 19 April 2021 at https://warwick.ac.uk/newsandevents/pressreleases/uber_drivers_are/.

International Labour Organization (ILO) (2012), 'Giving globalisation a human face: general survey on the fundamental conventions concerning rights at work in light of the ILO Declaration on Social Justice for a Fair Globalisation, 2008', International Labour Conference, 101st Session, Report III (Part 1B), Geneva.

International Labour Organization (ILO) (2018), *Freedom of Association: Compilation of Decisions of the Committee on Freedom of Association*, Geneva: International Labour Office.

Irish Competition Authority (2004), Case COM/14/03. 'Decision of the Competition Authority No. E/04/002 Agreements between Irish Actors' Equity SIPTU and the Institute of Advertising Practitioners in Ireland concerning the terms and conditions under which advertising agencies will hire actors', 31 August, Dublin.

Johnson, B. (2012), 'Developing legislative protection for owner drivers in Australia: the long road to regulatory best practice', in J. Fudge, S. McCrystal and K. Sankaran (eds), *Challenging the Legal Boundaries of Work Regulation*, Oxford: Hart, pp. 121–38.

Johnston, H. and C. Land (2018), *Organising On-demand: Representation, Voice and Collective Bargaining in the Gig Economy*, Geneva: International Labour Office.

Johnstone, R., S. McCrystal, I. Nossar, M. Quinlan, M. Rawling and J. Riley (2012), *Beyond Employment: The Legal Regulation of Work Relationships*, Sydney: Federation Press.

Jones, B. and J. Prassl (2017), 'The concept of "employee": the position in the UK', in B. Waas and G.H. van Voss (eds), *Restatement of Labour Law in Europe: The Concept of the Employee*, Oxford: Hart, pp. 747–70.

Kaufman, B. (2004), 'What unions do: insights from economic theory', *Journal of Labor Research*, **25** (3), 351–82.

Kellerson, H. (1998), 'The ILO declaration of 1998 on fundamental principles and rights: a challenge for the future?', *International Labour Review*, **137** (2), 223–8.

Kessler, F. (2017), 'The concept of "employee": the position in France', in B. Waas and G.H. van Voss (eds), *Restatement of Labour Law in Europe: The Concept of the Employee*, Oxford: Hart, pp. 197–218.

Klare, K. (2000), 'Countervailing workers power as a regulatory strategy', in H. Collins, P. Davies and R. Rideout (eds), *The Legal Regulation of the Employment Relation*, London: Kluwer Law International, pp. 63–82.

Kovacic, W.E. and C. Shapiro (2000), 'Antitrust policy: a century of economic and legal thinking', *Journal of Economic Perspectives*, **14** (1), 43–60.

La Hovary, C. (2013), 'Showdown at the ILO? A historical perspective on the employers' group's challenge to the right to strike', *Industrial Law Journal*, **42** (4), 338–68.

Landa Zapirain, J.-P. (2012), 'Regulation of dependent self-employed workers in Spain: a regulatory framework for informal work?', in J. Fudge, S. McCrystal and K. Sankaran (eds), *Challenging the Legal Boundaries of Work Regulation*, Oxford: Hart, pp. 155–70.

Manning, A. (2003), *Monopsony in Motion: Imperfect Competition in Labor Markets*, Princeton, NJ.: Princeton University Press.

Marinescu, I. and H.J. Hovenkamp (2019), 'Anticompetitive mergers in labor markets', *Indiana Law Journal*, **94** (3), 1031–63.

Marinescu, I.E. and E.A. Posner (2019), 'Why has antitrust law failed workers?', *SSRN Electronic Journal*, 1–41, doi:10.2139/ssrn.3335174.

Masterman, C.J. (2016), 'The customer is not always right: balancing worker and customer welfare in antitrust law', *Vanderbilt Law Review*, **69** (5), 1387–422

McCrystal, S. (2007), 'Collective bargaining by independent contractors: challenges from labour law', *Australian Journal of Labour Law*, **20** (1), 1–28.

McCrystal, S. (2009), 'Is there a "public benefit" in improving working conditions for independent contractors? Collective bargaining and the Trade Practices Act 1974 (Cth)', *Federal Law Review*, **37** (2), 263–94.

McCrystal, S. (2014a), 'Collective bargaining beyond the boundaries of employment: a comparative analysis', *Melbourne University Law Review*, **37** (3), 662–98.

McCrystal, S. (2014b), 'Designing collective bargaining frameworks for self-employed workers: lessons from Australia and Canada', *International Journal of Comparative Labour Law and Industrial Relations*, **30** (2), 217–42.

McGaughey, E. (2019), 'Uber, the Taylor Review, mutuality and the duty not to misrepresent employment status', *Industrial Law Journal*, **48** (2), 180–98.

Menegatti, E. (2015), 'Mending the fissured workplace: the solutions provided by Italian law', *Comparative Labor Law Policy Journal*, **37** (1), 91–120.

Menegatti, E. (2019), 'Employment protection for workers in the gig economy: comparative perspectives', paper presented to a seminar at Sydney Law School, University of Sydney, 9 April.

Naidu, S., E.A. Posner and G. Weyl (2018), 'Antitrust remedies for labor market power', *Harvard Law Review*, **132** (December), 536–601.

Netherlands Competition Authority (2007), 'CAO-tariefbepalingen voor zelfstandigen en de Mededingingswet. Visiedocument van de Nederlandse Mededingingsautoriteit', Netherlands Competition Authority, The Hague.

Netherlands Competition Authority (2017), 'Leidraad. Tariefafspraken voor zzp'ers in cao's', Netherlands Competition Authority, The Hague.

Netherlands Competition Authority (2020), 'Leidraad. Tariefafspraken zzp'ers', Netherlands Competition Authority, The Hague.

Novitz, T. (2003), *International and European Protection of the Right to Strike*, Oxford: Oxford University Press.

Paul, S. (2016), 'The enduring ambiguities of antitrust liability for worker collective action', *Loyola University Chicago Law Journal*, **47** (3), 969–1048.

Peritz, R.J.R. (2000), *Competition Policy in America: History, Rhetoric, Law*, Oxford, Oxford University Press.

Perulli, A. (2003), *Economically Dependent/Quasi Subordinate (Parasubordinate) Employment: Legal, Social and Economic Aspects*, Brussels: European Commission.

Prassl, J. (2018), *Humans as a Service: The Promise and Perils of Work in the Gig Economy*, Oxford: Oxford University Press.

Prassl, J. and M. Risak (2016), 'Uber, Taskrabbit, & Co: platforms as employers? Rethinking the legal analysis of crowdwork', *Comparative Labor Law and Policy Journal*, **37** (3), 619–52.

Regulation 2016/679 of the European Parliament and of the Council of 27 April 2016 on the protection of natural persons with regard to the processing of personal data and on the free movement of such data, and repealing Directive 95/46/EC (General Data Protection Regulation), Official Journal of the EU L 119/1.

Roles, C. and A. Stewart (2012), 'The reach of labour regulation: tackling sham contracting', *Australian Journal of Labour Law*, **25** (3), 258–283.

Rosenblat, A. (2018), *Uberland: How Algorithms Are Rewriting the Rules of Work*, Oakland, CA: University of California Press.

Riley, J. (2005), 'Who owns human capital? A critical appraisal of legal techniques for capturing the value of work', Australian Journal of Labour Law, **18** (1), 1–25.

Sanchez Torres, E. (2010), 'The Spanish law on dependent self-employed workers: a new evolution in labor law', *Comparative Labor Law and Policy Journal*, **31** (2), 231–48.

Schlachter, M. (2019), 'Trade union representation for new forms of employment', *European Labour Law Journal*, **10** (3), 229–39.

Slee, T. (2017), *What's Yours is Mine: Against the Sharing Economy*, London: OR Books.

Staiger, D.O., J. Spetz and C.S. Phibbs (2010), 'Is there monopsony in the labor market? Evidence from a natural experiment', *Journal of Labor Economics*, **28** (2), 211–36.

State of Victoria (2020), *Report of the Inquiry into the Victorian On-demand Workforce*, Melbourne: Victorian Government.

Steinbaum, M. (2019), 'Antitrust, the gig economy and labor market power', paper presented to the Duke *Journal of Law and Contemporary Problems* Labor Symposium, Duke University, Durham, NC, March.

Stewart, A. (2002), 'Redefining employment: meeting the challenge of contract and agency labour', *Australian Journal of Labour Law*, **15** (3), 235–76.

Stewart, A. and S. McCrystal (2019), 'Labour regulation and the great divide: does the gig economy require a new category of worker?', *Australian Journal of Labour Law*, **32** (1), 4–22.

Stone, K. (2004), *From Widgets to Digits: Employment Regulation for the Changing Workplace*, Cambridge: Cambridge University Press.

Stone, K. and H. Arthurs (eds) (2013), *Rethinking Workplace Regulation: Beyond the Standard Contract of Employment*, New York: Russell Sage Foundation.

Todoli-Signes, A. (2019), 'Workers, the self-employed and TRADEs: conceptualisation and collective rights in Spain', *European Labour Law Journal*, **10** (3), 254–70.

US Department of Justice and US Fair Trade Commission (2016), 'Antitrust guidance for human resource professionals', accessed 19 April 2021 at https://www.justice.gov/atr/file/903511/download.

US Department of the Treasury (2016), 'Non-compete contracts: economic effects and policy implications', report, accessed 19 April 2021 at https://www.treasury.gov/resource-center/economic-policy/Documents/UST%20Non-competes%20Report.pdf.

US White House (2016), 'Non-compete agreements: analysis of the usage, potential issues, and state responses', report, accessed 19 April 2021 at https://obamawhitehouse.archives.gov/sites/default/files/non-competes_report_final2.pdf.

Waas, B. and G. Heerma van Voss (2017), *Restatement of Labour Law in Europe: The Concept of the Employee*, vol. 1. Oxford: Hart.

Wedderburn, Lord (1986), *The Worker and the Law*, 3rd edn, Harmondsworth: Penguin Books.

Weil, D. (2017), *The Fissured Workplace: Why Work Became so Bad for so Many and What Can Be Done to Improve It*, Cambridge, MA: Harvard University Press.

Whiteside, N. (2019), 'State policy and employment regulation in Britain: an historical perspective', *International Journal of Comparative Labour Law and Industrial Relations*, **35** (3), 379–400.

PART II

Unpacking platform economy puzzles
– economic and social exchanges in
platform-mediated gig work

5.   Platform urbanism and infrastructural surplus

Aaron Shapiro

INFRASTRUCTURAL SURPLUS AND THE URBAN PUZZLE

Online platforms are transforming the social and economic relationships that organise everyday urban life. Millions, probably billions, of city-dwellers across the world now depend on platforms for mundane activities, and it is easy to understand why. Service platforms are attractive to consumers; they offer convenience, speed and accessibility at low costs. Online hubs for transportation, groceries, socialising, handiwork, dating, parking, healthcare, education and more have already disrupted nearly every corner of cities' economy and society, cutting through the ossified institutions and regulatory barriers of the post-war social contract (Huws 2016).

What about the production side? What is it about urban life that makes cities such an attractive source of value for technology giants, including publicly traded companies such as Amazon and Alphabet, firms whose valuations are among the highest in the world (Sadowski 2020b)? What role does the city play in the political economy of platform capitalism (Langley and Leyshon 2017; Srnicek 2016)?

The question may at first seem straightforward. After all, as Jathan Sadowski (2020a, p. 450) notes, platforms concentrate in cities 'for many of the same reasons that capital is centralised in cities', namely, the agglomeration effects of market concentration, consumer density and the availability of cheap labour. While those are certainly characteristics of urban centres, they are surely not all a city is. Cities are also arenas for, and objects of, political protest; the inspiration and canvas for cultural production; places where progressive, sometimes radical ideals become experiments in new ways of living together. The consequences of platforms' entrance into urban economies for the socially produced and collective value of urban life remains an ongoing empirical question.

This is precisely the point made by recent scholarship on 'platform urbanism' (cf. Barns 2018a, 2018b, 2019; Leszczynski 2020; Rodgers and Moore 2018). As a theoretical agenda, platform urbanism is concerned with the ways that platforms 'remediate forms of locational or urban value' (Barns 2018a). Instead of being regarded purely as political-economic entities, platforms are viewed as resources for consumers and governments grappling with long-standing regulatory challenges and failures (Davidson and Infranca 2015). The platform-urbanist approach emphasises the creativity with which urban subjects appropriate platforms. It advances a more hopeful vision in which technologies might support, not undermine, cities' value to their communities (Leszczynski 2020), since the richness of urban life can never be reduced to the 'transactional logics' in which platforms trade (Barns 2018a).

The trouble is that this hopeful vision hardly squares with the lived experiences of platform *workers*. Transactional logics have immediate consequences for platform labour. While some firms may frame their workers as partners (Rosenblat 2018), the realities of platform labour reveal deep asymmetries in power and resources between the platform and its workforce, and these asymmetries materialise in the platform-mediated transaction. Platforms do not value workers as creative contributors to the company's operations. Workers are human supports to platform-mediated networks, exchanges and amenities – what Prassl (2018) aptly terms 'humans as a service'. From workers' perspective, platform urbanism could just as easily describe the politics of the transaction: a struggle over the forms of work that platforms recognise as legitimate, compensable labour. How is the transaction defined? Who gets to set the terms, to determine where the legitimate, compensable work of platform labour ends? For example, as Van Doorn and Badger argue in Chapter 6 of this volume, platforms extract immense amounts of data about the transactions they facilitate. Platform work therefore produces not only an economic surplus for the platform in the form of transactional rent (that is, the platform's cut of consumer payments), but also a data surplus, an 'informational service' that amplifies platforms' financial and technical power but for which workers are paid nothing (see also, Attoh et al. 2019). Similarly to Van Doorn and Badger, I am troubled by this duplicity. However, and similar to the platform urbanists, I am interested in the outside to the transactional data, in the ways that transactions reduce the embodied, intimate and relational experiences of the 'technological everyday' (Amin 2011, p. 109) to computationally legible signals. What does transactional data conceal about working on the platform and in the city?

Is it possible to take the lessons of platform urbanism and insist on the irreducibility of urban life, but with an eye toward the platform-mediated transactions that capitalise on cities' socially produced value (Rossi 2019)? This is the urban puzzle that I tackle in this chapter. The solution that I propose lies in platforms' ability to capture what I term 'infrastructural surplus'. Infrastructural surplus denotes the excess of value derived from collective resources embedded in the urbanised landscape. In some instances, platforms appropriate this value directly (think of the roads and bike lanes traversed by Uber drivers and Deliveroo riders, for example). In others, however, platforms capture infrastructural surplus indirectly, through their workers, whose local knowledge of, and skills in, navigating urban resources become vital assets under increasingly precarious platform labour conditions, particularly in the invisible work of caring for and maintaining their bodies and equipment (Casilli 2017; Daniels 1987; DeVault 2014; Jackson 2014; Mattern 2018; Rossi 2019; Star and Strauss 1999).

I elaborate on the theoretical dimensions to infrastructural surplus in the next section by connecting the invisible work of platform labour to urban infrastructures as collective consumption goods. I then offer an anatomy of infrastructural surplus and consider two modes by which platforms extract value from collective goods: reformatting social space and transactional exclusion. I conclude by discussing the implications of infrastructural surplus for policy and regulation, and suggesting directions for further research.

THEORETICAL GAP: THE INVISIBLE WORK OF URBAN INFRASTRUCTURE

According to Sarah Barns, a leading voice in the burgeoning platform urbanist tradition, the platformisation of urban life always 'encompasses, but also exceeds, the transactional data of urban interactions' (Barns 2019, p. 7). What is this excess to transactional data? For Barns and other platform urbanists (for example, Leszczynski 2020), it consists of the mundane activities of everyday life: socialisation, commuting, civic engagement, cultural expression, and so on. 'Platform urbanism, enacted daily as we commute, transact, love, post, listen, tweet, or chat, deeply implicates the everyday urban encounter' (Barns 2018b). That we consume platforms in particular places, while engaging in routine activities, territorialises platform sociality in mundane urban experiences. It brings the abstracting data of tags and likes and clicks and views down from the cloud and onto city streets, where engagement takes on a particular valence and significance in the lives of urban subjects. The analysis therefore centres on the contextual richness that gets lost in platforms' intermediations – when a teenager watches YouTube videos on a crowded subway on the way home from school, or a commuter decides to order an Uber on a particularly wet and cold day, or when I comment on a friend's social media post from a public park.

That is, platform urbanism's primary concern centres on how urban subjects consume platforms in cities. It tells us very little about the labour or production side of platform economies. For this we need a different grasp on the platform transaction and its excess, not the rich, contextual details of everyday life that the platform cannot see, but all the factors that the platform refuses to see and that it strategically excludes from the transaction to minimise costs, liabilities and obligations as an employer.

All work exceeds the 'grammars of action' that employers recognise in labour transactions (Agre 1994), and this excess is particularly salient in work mediated by computers (Star and Strauss 1999). As a mobilisation of computational mediation into everyday urban transactions, platforms such as Uber (ride-hailing) and Deliveroo (restaurant delivery) codify those grammars in the architecture of the worker-facing app. What counts as labour are only those actions prefigured as measurable and thus legible to the platform. Through this reductive datafication (van Dijck 2014), platforms condense the qualitative experience of work into a quantitative frame, pre-emptively excluding the mess, complexity and intensity of labour in urban environments (Barns 2019; Briziarelli 2018). For example, Instacart refused its delivery workers hazard pay during the COVID-19 pandemic, denying that health risks are relevant to the labour transaction; when the Seattle City Council passed legislation requiring an additional \$5 per order, the company threatened to cease local operations (Nickelsburg 2020). Although an extreme example, this illustrates the stakes of labour visibility and recognition and, by extension, invisibility and misrecognition (cf. Van Doorn 2017). The excess to transactional data is not only the urban context that the platform cannot see, but all the work and costs that it refuses to recognise.

Brian Massumi describes the arbitrage of (in)visibility and (mis)recognition as the market's 'immanent outside' – 'factors that belong to capitalism's field but do not belong to its system' (Massumi 2018, p. 9). The outside to platform work comprises factors that belong to the platform's field but which are not recognised by its system; bike couriers' need to stay hydrated on a hot day, for example, or to keep their smartphones powered to continue receiving delivery instructions, or to find a toilet in cities without public facilities. Platform workers must learn these mundane, invisible but no less essential practices, skills and knowledge on their own, and to do so for free, outside the transaction. It is therefore unsurprising that some platform workers develop an intuitive sense of the transaction's limits, learning to view technical interventions such as app updates cynically, if not with outright suspicion (Shapiro 2018).

The (in)visibility of work is an enduring concern in the politics of labour, especially in societies where our willingness to see work maps onto racialised and gendered divisions (Van Doorn 2017). Feminist critiques of Marxism have long decried that economic institutions, cultural norms and legal regimes recognise only particular types of work as productive, whereas the reproductive work of maintenance, care and repair remained invisible, and usually uncompensated (Casilli 2017; Daniels 1987; DeVault 2014; Jackson 2014; Star and Strauss 1999; Weeks 2007). Reproductive work sustains the labour force: it is what keeps productive workers fed, rested and clothed. Historically coded as feminine, reproductive work involves an immense amount of time and energy – everything from education, civic engagement and socialisation to nutrition, clothing, shelter and healthcare; that is, all the aspects of life that employers expect to take place outside the workplace but without which productive labour would be impossible. What will be seen as legitimate labour is therefore always a contested issue, mediated by context-dependent regimes of visibility. As Star and Strauss (1999, p. 9) argue, 'No work is inherently either visible or invisible. We always "see" work through a selection of indicators… The indicators change with context, and that context becomes a negotiation about the relationship between visible and invisible work.'

Platform labour is unique in at least two ways. First, the classification of platform workers as independent contractors (instead of employees) rewrites the post-war labour contract, tilting the balance of visible and invisibilised work towards the latter and reversing hard-won gains in employee protections, most immediately, because contractors are expected to provide and maintain their own equipment and resources (de Stefano 2015; Huws 2016). Second, platforms offer far fewer opportunities for negotiating the productive-visible/reproductive-invisible divide than a human manager might. As Lianos and Douglas (2000, p. 264) suggest, negotiating with automated sociotechnical environments (of which platforms are only the most recent and perhaps most profitable example) 'is by definition impossible'. The technology firms that produce platforms exercise their authority by defining the indicators through which we see work. They determine which actions will be visible and recognised as legitimate labour. They dictate where productive work ends and where the offline work of reproduction begins; for example, Uber drivers compelled to clean their vehicles' interiors after logging off, or timing their shifts to avoid stopping in the middle of a ride for fuel. These and other offline tasks are the sole responsibility of the entrepreneurial gig worker.

What differentiates urban platform economies is that they ground these time-tested strategies of misrecognition and invisibility in the spatial politics of the urban realm (Briziarelli 2018; Rossi 2019). As with freelancing and creative work before it (Gregg 2011), gig work disperses the labour context into everyday spaces of consumption and reproduction. Work takes place on streets and sidewalks, and in and between restaurants, shopping centres, grocery stores and apartment complexes. The city – more precisely, the urbanised landscape (Amin and Thrift 2002; Brenner 2013) – is gig work's 'factory' (Greenberg and Lewis 2017). Workers must therefore deal with a host of frustrations that come with working in the city, with navigating the urbanised landscape; congestions, delays, stoppages, and so on, all barriers that on the piecemeal or commission wage systems that most platforms use conspire to decrease workers' hourly take-home earnings (Shapiro 2018).

Similarly, the urbanised landscapes where platform work takes place is also brimming with resources; for example, systems, networks, institutions and commercial spaces that privilege accessibility and collective use (Richardson 2018). Manuel Castells (1973) famously defined urbanisation as the spatial concentration of those resources, or what he termed collective consumption goods: the means and infrastructures by which urban populations reproduce themselves and their capacity to work. Collective consumption goods include physical infrastructures for housing, transportation, energy and natural space, as well as social infrastructures for education, childcare, entertainment, healthcare, and so on. As Herman and Ausubel (1988, p. 1) write:

> Cities are the summation and densest expressions of infrastructure, or more accurately a set of infrastructures, working sometimes in harmony, sometimes with frustrating discord, to provide us with shelter, contact, energy, water and means to meet other human needs ... The physical infrastructure consists of various structures, buildings, pipes, roads, rail, bridges, tunnels and wires. Equally important and subject to change is the 'software' for the physical infrastructure, all the formal and informal rules for the operation of the systems.

As platforms expand their operations in the urbanised landscape, these collective infrastructures become vital assets to workers under increasingly precarious labour conditions. What I hope to demonstrate in the following pages, however, is that platforms are also capitalising on these infrastructures, directly, by grounding their operations in existing material and labour networks for which they pay nothing, and indirectly, through their arbitration of the labour transaction.

The process mimics how industrial capitalists profit from workers' labour, through the appropriation and accumulation of surplus. However, whereas the classic Marxist scene of exploitation plays out as an arbitrage of labour- and exchange-value in the commodity, here surplus derives from the inexhaustible utility and accessibility of infrastructure – from infrastructure's publicness: its collective utility, its non-monetary surplus value (Teubner 2020). As Lizzie Richardson (2018) argues, platforms' explosion into the urban lifeworld heralds 'a condition in which the public and private qualities of urban infrastructures are being reconfigured'. Within this condition, the publicness of infrastructures is less a question of state or corporate ownership than a 'quality of accessibility for collective use'. Platforms exploit that accessibility of use by policing the boundaries of the labour transaction to exclude, first, the fixed capital invested in collective goods and, second, workers' knowledge and ability to navigate public infrastructures.

This latter exclusion, in particular, deserves greater scrutiny. The invisible, infrastructural work of reproduction becomes increasingly vital as the precaritisation of gig work intensifies. Lacking even a modicum of support from the platform, public infrastructures are critical to the work of caring for and maintaining workers' bodies and equipment, which the platform demands but for which it pays nothing. To make it in platform economies, workers must develop a critical 'gig literacy' and 'algorithmic competencies' (Jarrahi and Sutherland 2019; Sutherland et al. 2020). In urbanised platform economies, that literacy also involves an embodied, dispositional acuity in infrastructural resources; that is, an urban habitus (Bourdieu 1977) that fills the infrastructural gaps of the lean business model popularised by Uber and later adopted by so many technology start-ups (Srnicek, 2016). Indeed, aside from intellectual property claims (Birch 2019), technology start-ups own few assets. They pride themselves on their leanness, for instance, by boasting of high ratios of outsourced contractors to internal employees (for example, Milbourn 2015). As Richardson (2018) notes of Uber, the list of assets that platforms own is slim: 'The roads, the cars, and even many of the drivers are already part of an [existing] urban infrastructure. In fact, the single novel element appears to be the Uber application, a piece of software'. Similar to other service platforms, Uber does not maintain the roads or its fleets of cars, nor does it train and certify its workers. The value of these skills, knowledge and resources, coming together in the platform-mediated transaction (Richardson 2020), enables the capture of infrastructure's surplus, the socially produced value of infrastructure's publicness.

Infrastructural surplus reveals the underbelly of platform urbanism, not how urban subjects make use of platforms in everyday life, but how platforms appropriate urban subjects and infrastructures as supports to their own lean, extractive operations.

CAUSES AND CONSEQUENCES OF THE PROBLEM: TOWARDS AN ANATOMY OF INFRASTRUCTURAL APPROPRIATION

Infrastructural surplus provides a novel perspective on the role of urbanisation in the political economies of platform capitalism. As critical urban theorists have emphasised in recent decades, the urban emerges through the density of infrastructural entanglements that constitute the built environment (Amin 2014; Amin and Thrift 2002; Gandy 2005; Graham and Marvin 2001). Extending this premise to platform economies, we might say that platforms become urban by grafting their extractive operations onto existing infrastructural systems and entanglements. To specify the process, this section proposes an anatomy to delineate the means by which platforms appropriate surplus from infrastructural resources. It then proceeds to identify two modes of extraction to illustrate the role of infrastructural surplus in platforms' extractive operations: what I describe as 'reformatting social space' and 'transactional exclusion'.

The first and most primitive form of appropriation involves the physical systems, networks and devices that allow for movement in and through the built environment; that is, the type of objects most immediately associated with the term 'infrastructure': 'artifacts built of concrete and steel: the "hard" technical systems that facilitate the distribution of people, energy, water, waste, information' (Carse 2012, p. 4). One thinks immediately of the roads that ride-hailing drivers on platforms such as Uber and Lyft traverse, or the bike lanes used by restaurant delivery riders on platforms such as Caviar or Deliveroo. However, the category might also include any of the myriad material, spatial and distributive elements that furnish urban life and lend support, unwittingly, to platform operations: cell phone lots at airports where Uber drivers wait for ride-requests (Wells et al. 2020); parks and public squares where delivery riders congregate between jobs (Briziarelli 2018); the homes that Airbnb hosts rent out at higher rates owing to their proximity to cultural landmarks or central business districts. Then, too, there are the means of transportation that workers bring with them to the job: Uber drivers' cars and Deliveroo couriers' bikes. All of these represent essential and often invisible nodes of the infrastructural assemblage that platforms enact, assets whose maintenance and associated costs are externalised onto municipal governments, individual homeowners, airport authorities and, not least, workers. Platforms thrive off of the liveliness that these infrastructures make possible (Amin 2014) but contribute little, if anything to their upkeep.

A second level of appropriation involves network infrastructures. As ample reporting in the popular press has rhapsodised, the rise of platform economics follows from sweeping technological advances, the most notable being the smartphone – the clarion of mobile computing's manifest destiny (for example, The Economist 2014; see also Miller 2014). Both platform consumption and platform labour require some degree of connectivity, whether through Wi-Fi or 5G (fifth generation cellular data infrastructure). As Barns (2018b) puts it, platform urbanism begins in the 'everyday interactions of smartphone-equipped urban subjects'. Behind the veneer of this wireless interactivity, however, lies an intensely material network of cell towers and wireless routers, in addition to the protocols that govern data transmission and exchange (Galloway 2004; Mackenzie 2005). There is overlap with the physical systems discussed in the previous paragraph, only now in reference to the systems maintained by telecommunications conglomerates, their subsidiaries and other network providers. However, connectivity requires a linkage between this distributed material and protocological infrastructure to the bodies of consumers and workers; hence the need for smartphones. As former

Uber employee Scott Gorlick explained in a series of Twitter posts, Uber's ability to scale from two cities to hundreds of urban markets across the world required rapid recruitment campaigns. In early days, this involved 'literally [giving] out iPhones and accessories to drivers' (Gorlick 2020). The platform could not just recruit a workforce; it had to equip its new workers as nodes in a connective network. The sunk costs of this equipment paid off as Uber came to dominate the ride-hailing sector and workers were left covering the costs of the smartphones' data plans.

A third register involves platforms tapping into cities' existing labour infrastructures. Gorlick's reflections on Uber's growth again illustrate the point: 'When we launched a city, we'd go on Yelp and get a list of all limo companies in the area. We would cold call, meet and recruit drivers to join … Limo and taxi drivers spent a lot of time waiting in parking lots around the airport. So we spent a lot of time down at the airport' (Gorlick 2020). Here we see plainly that Uber recruited its workforce, first, by tapping into taxi and limousine companies' driver bases. Once that labour pool was saturated, the platform then began incentivising drivers to recruit their friends: 'Power drivers (who did lots of trips) referred people who were likely to become power drivers' (Gorlick 2020). However, most importantly, Uber – flush with venture funding – actively sabotaged its industry competitors from which it was sourcing labour by under-pricing the platform service and decreasing costs of entry: 'If you drove for a limo/taxi co, you'd keep only 30% of your fares or have to rent a car for $500/week. Drivers left limo cos, bought cars & started making more money w Uber' (Gorlick 2020). This honeymoon period would not last long, though; once the workforce was large enough, Uber quietly began decreasing its baseline pay, decoupling drivers' earnings from what passengers paid for a trip (Rosenblat 2018, p. 108). Drivers' earnings plummeted, but many found themselves trapped in a cycle of debt perpetuated in some instances by Uber's now-defunct subprime auto-loan programme (Leberstein 2016). My point is not that the taxi or limousine industries were any less exploitative than the platform model (cf. Mathew 2008), only that Uber used its financial resources to recruit from an existing labour infrastructure in a way that: (1) sabotaged industry competitors, (2) avoided existing regulatory barriers – driver training, certification programmes and licensing, for example – and then (3) chipped away at workers' earnings by constantly redefining the terms of the labour transaction (Rosenblat 2018).

These three registers of infrastructural appropriation are not exhaustive, nor are they mutually exclusive. Similar to an anatomy, they may be parsed and examined as discrete specimens, but must be understood as always working in concert. As Rossiter (2016) insists, it is the imbrications across networked software systems, labour regimes and infrastructures that form the crux of contemporary capitalism's operational modality (see also, Mezzadra and Neilson

2017). The remainder of this section identifies two modes of extraction that cut across the different registers of infrastructural appropriation. The first involves platforms reformatting social space as an infrastructural support for their operations; the second excludes reproductive work from the platform-mediated labour transaction.

Reformatting Social Space: The Case of Instacart

The on-demand grocery-delivery start-up Instacart has received abundant praise from the business press as an innovator (Barry 2016; Kolodny 2016). In 2015, for example, Fast Company included Instacart in its list of retail's 'most innovative companies' (Fast Company 2015). Yet it seems little is novel about the platform's business proposition. Companies such as Kozmo and Webvan pioneered online grocery delivery in 1990s, only to later become famous stand-ins for the failed speculations of the dotcom era. Both Kozmo and Webvan flamed out: they burned through stockpiles of venture capital funds or proceeds from initial public offerings (IPOs) as they attempted to scale the logistics of online grocery delivery (Bensinger 2015; Etherington 2013). What differentiated Instacart, that is, its innovation, was therefore operational. Whereas Webvan acquired warehouses to stock its own inventory, Instacart does not sell products at all. The platform hires personal shoppers (a combination of part-time employees and contract workers) to collect and purchase items from existing supermarket chains, check-out with prepaid debit cards and then deliver those items to customers' doorsteps. Whereas Webvan was a grocer that delivered, Instacart claims to be a platform for connecting consumers to 'third party logistics providers' who shop on their behalf (Instacart 2019).

Sociologist Chelsea Wahl conducted ethnographic research as an in-store personal shopper with Instacart, and her findings reveal parallels between working the aisles of a supermarket fulfilling customers' orders and the breakneck labour of picking at Amazon's fulfilment centres, perhaps currently the quintessential scene of logistical labour (Wahl in progress). Instacart's revenues come from the delivery fee that it charges customers; the more orders its shoppers complete, the more revenue the company collects. The platform therefore incentivises speed by paying workers based on a commission system, combining per-item and per-order rates (Whitney 2017). The platform also uses adaptive metrics to benchmark worker performance. In combination, commissions and metrics induce a constant pressure for workers to locate, scan, weigh and assemble items on the grocery list as efficiently as possible (Wahl in progress). As with the devices that direct and track Amazon pickers through its nearly million-square-foot fulfilment centres (see Guendelsberger 2019), workers' smartphones display an anxiety-inducing countdown of the time allotted to gather each item on the customer's grocery list. The difference is that Amazon's optimised inventory is not designed to be knowable or even legible to humans (cf. Bowles 2020), whereas Instacart shoppers must learn and memorise the layout of the supermarket, streamlining their movements to conform to the platform's logistical demands (cf. Kanngieser 2013).

Food writer Alyse Whitney (2017) reported her experience working for a day as an Instacart shopper. In Whitney's experience, learning the store's layout was a first-order priority; that knowledge clearly distinguished veteran workers' skills from her own novice performance. Equally important, Whitney found that she had to shed her sensibility as a consumer-shopper and adopt the more calculating, efficient routine of an Instacart 'shopper':

> First off, I didn't know the layout of this store. Other Instacart employees literally ran circles around me as I searched for a very specific brand of almond milk. 'Just pick another unsweetened vanilla,' Jean suggested, but as someone who also has brand loyalty to Califia Farms, I opted to shoot a quick message to the customer to approve a plain vanilla of the same brand instead. She approved a few minutes later, and I was onto the next one.

Being quick to identify replacement items, and avoiding unnecessary and time-consuming communications with the customer, is a learned component of the Instacart worker's habitus (Bourdieu 1977). The bodily routine of being a shopper requires adaptation, and the platform structures the labour protocols to ensure compliance. When workers check out at the register, for instance, the prepaid debit cards are rejected if the total does not correspond to the expected cost. To pre-empt such rejections, workers scan every item on the list to ensure the correct price and quantity. For produce, this means weighing every item, down to the last ounce or gram. Before some helpful advice from another veteran Instacart shopper, Whitney was scanning items and weighing produce individually. This was not how a shopper shopped; best to load the cart with items and then scan and weigh in bulk, she was told. Finally, while all the scanning, messaging, and weighing eats into the workers' allotted time, it also burns through the batteries that power workers' smartphones. Whitney therefore found herself scrambling to keep her device energised between trips through the aisles.

Conforming to these efficiency demands means workers must modify their habits and movements, the ways they engage with the supermarket as a consumer space. As Farhad Manjoo (2014) reported in the *New York Times*, it is astonishing to realise 'how unexpectedly difficult it is to quickly and accurately buy and deliver an assortment of groceries for a stranger'. This already speaks to the invisible work of platform labour. However, we can also imagine the scene of Instacart's operations as a microcosm for the spatial politics of the platform economy writ large. Platform innovation, glossed through the prism of infrastructural surplus, requires repurposing consumer spaces; in this example, spaces that are maintained and restocked by some entity external to the platform and its workforce. Whereas in-store customers will continue to experience the grocery store as a space of consumption, workers experience the same space as a logistical scene: a warehouse, its aisles now rows of inventory, and its staff middlemen in a supply chain.

Supermarket operators have been ready and willing to accommodate the logistical overhauls necessary to support Instacart's operations. Since the platform benefits supermarkets by increasing sales, many of Instacart's supermarket partners add capacity to accommodate its operational needs. As Manjoo (2015) explains, partnerships helped the upstart platform 'improve efficiency' even more in its earliest days. In the most popular outlets, once Instacart pickers (Manjoo's term) have gathered and paid for a customer's order, they can then store the groceries in dedicated shelves and refrigerators for delivery drivers to pick up more easily. Many stores also set aside Instacart-only cash registers so the shoppers can avoid lines. Finally, Instacart builds software for retailers to integrate their inventory with the platform's online system and, since 2015, has developed 'machine-learning tools to predict inventory shortfalls and the number of staff members it will need in each store throughout the day' (Manjoo 2015; see McCreight 2019).

This ad hoc assemblage of software (machine learning tools and the worker-facing app), infrastructure (the supermarket-as-warehouse) and labour (shoppers, cashiers, and so on) was put to the test during the COVID-19 pandemic. With municipalities enacting shelter-in-place orders, Instacart experienced a 500 per cent increase in daily order volume; this dramatic spike in demand put a strain on what Instacart's executives call its just-in-time operational model (Schaaf 2020). Anecdotally, I observed grocery outlets in my city filling the gaps by increasing the capacity allotted to Instacart's operations. At one market, a large and once popular café at the building's entrance was replaced with alphanumerically coded shelves to hold Instacart orders awaiting pick-up by delivery drivers. This space was needed to make room for the 250 000 new shoppers that the platform brought on board in April 2020 (Mascarenhas 2020). On the back end, Instacart's engineers and computer scientists scrambled to improve their software to keep up with inventory shortages and out-of-stock notifications from partner stores (Schaaf 2020). Through it all, however, the model remained fundamentally the same: Instacart exploits the supermarket's accessibility and collective utility – its publicness – to serve as an infrastructural base for its operations.

Transactional Exclusions: The Case of Caviar

The spatial politics of logistical labour become far messier when we move outside the orderly aisles of the supermarket and onto city streets. This is where the work of restaurant delivery platforms takes place: in and between restaurants, alleyways, service entrances, high-rise apartment buildings, plazas and the securitised lobbies of office buildings. For workers on these platforms, the workplace is the urbanised landscape itself (Wells et al. 2020). Couriers have to deal with traffic jams and rainstorms; as cyclists, they are exposed to bodily harm, accidents and exhaustion. The city is not just a context for this work: shuttling through urban space quickly and safely is the job.

Delivery platforms are therefore exemplary of a second mode of infra-structural appropriation: transactional exclusion. Whereas Instacart's value proposition stems from the reformatting of a consumer space into a logistical infrastructure, delivery platforms such as Caviar capture infrastructural surplus indirectly, through their arbitration of the labour contract, by excluding the reproductive work necessary for the job. The 'grammars of action' (Agre 1994) involved in running deliveries, that is, getting to a restaurant, picking up the order and then delivering it to the customer, is formalised in the step-wise prompts that workers must complete on the worker-facing app (Shapiro 2018). However, there is a raft of work that precedes and exceeds this simple sequence, especially when it comes to bodily care and equipment maintenance. Under precarious labour conditions, public infrastructures embedded in the urban environment become critical resources, especially in instances of break-down, malfunction and disrepair.

Miranda's experience is exemplary.[1] When I interviewed her in 2016, Miranda had been working as a bicycle courier for Caviar in Philadelphia for about a year. She was generally happy with the situation, but as we talked, she told me of a number of experiences that caused her to doubt whether delivering for Caviar was really 'worth it'. One instance involved a flat tyre. Miranda was 'pretty good' about bike maintenance; her boyfriend had worked as a bike mechanic and helped her when issues came up. She put air in the tyres weekly, sometimes more often, and she regularly cleaned her bike chain with fresh lubricant. Even so, everyone gets a flat tyre on occasion. In this instance, she happened to notice the flat tyre on her way to a customer's home after having already picked up a delivery from a restaurant. Miranda used a feature on the Caviar app to message a remote dispatching service and inform them of the sit-uation. A dispatcher responded within a few minutes that another courier was

---

[1] The interviews were conducted in Philadelphia, PA, from December 2015 to May 2016. Workers' names have been changed to protect their identity.

on the way to intercept and complete the delivery to the customer. The hand-off went smoothly, and Miranda was able to patch the flat tyre; she even took a few more orders before logging off for the day. However, when she checked the record of her daily activity, Miranda noticed that she was never paid for the order that she handed off to another courier. She contacted Caviar's help team and was told that the other courier must have received payment for the order, despite the fact that he only worked the delivery's final stretch. The help team sent Miranda a bonus of $5 to make up for the lost earnings, but the delivery would have paid out $12. A few months later Miranda had another bike issue ('I don't remember – the chain got jammed or something'). This time, rather than report it to the dispatcher, she locked her bike to a street sign and rented a bike from a nearby bike-share station, paying $4 for a half-hour's worth of use. 'It just made more sense – you know, economically – than someone else getting paid for my order'.

None of the work that Miranda puts into maintaining her bike is legible to the platform. This is the invisible work of repair and sociotechnical reproduction that all platform workers – but especially bicycle couriers – are expected to undertake. Running deliveries by bicycle also requires a certain level of physical fitness that other types of work do not, and thus involves a number of proactive steps to ensure well-being (carrying extra water, eating carbohydrate-dense meals before a shift, and so on). Moreover, cyclists are directly exposed to the elements. This creates an additional onus in adverse conditions, and workers accumulate piles of gear for biking in the rain, sleet, sun and snow. All of this work is a mutual expectation. Many couriers enjoy the challenge and exercise, and mechanical skills are a point of pride among an older generation of bike workers (see Stehlin 2019).

What I found so striking about Miranda's anecdote was the agility and nonchalance with which she navigated the interruptions, for example, the tacit knowledge of bike-share locations as an economical backstop in emergency situations, and that she repaired the flat tyre and continued riding. All of these actions are invisible to the platform, and they speak to her embodied dispositions, as both urban subject and platform worker. She navigates hazards on city streets daily, and she navigates the precarity of piecemeal wages and capricious decision-making on the platform. While this is just one story in the universe of gig tales (many of which are shared online on forums such as Reddit), the anecdote gets at a feature fundamental to platform work, namely, that it is perhaps the platform, more than workers, which benefits from workers' dexterity in making use of urban infrastructures as a resource. Similar to the Instacart shopper conforming to the platform's efficiency demands, couriers take countless steps to avoid interruptions, to move between restaurant and customer as efficiently as possible. Also, if we envision this invisible work scaling up to the size of an entire courier fleet, that local knowledge and embodied expertise begins to function like the lubricant on a bike chain – greasing the cogs of the platform's operations.

Some tactics can be extreme. I interviewed a courier who once delivered a meal by subway because he figured it would be faster than biking the same distance on a cold, rainy night; another claims that he has paid, out of pocket, dozens of parking tickets that he received while making deliveries by car (Caviar pays one ticket for every hundred orders completed). Most often, it is the small, mundane acts of reproductive work that structure the experience of working the city, and which connect platform labour to the circuitry of urban infrastructure. For example, I asked Mike, a veteran courier with long dreadlocks, how he stays hydrated:

> Alright, so I actually have a good system for this. I mean, there's always the restaurants. You're in there, waiting for your order, you can just go ask someone and they usually just let you fill up from the sink or whatever. But if you just finished a delivery and you're really thirsty, there's also spots – but I guess it depends on where you are. Like, let's say you're near the Schuylkill [a river running through Philadelphia with an adjacent trail]. There's these water fountains down there that actually have, like, [a spout] for filling up water bottles. Those are good … But actually, now that I think about it, I really like getting water from Wawa [a local chain of convenience stores] because you can just walk in and go right to the soda machine and press the water button – they'll never say anything to you. It's just – I don't know, it's just really cold or fresh or something, but that water tastes really good, especially when it's hot out. Also, most Wawas have a bathroom too, so that's always good. You just kind of memorise where the Wawas are.

To paraphrase Mike: workers map a complex network of resources, from public fountains to restaurants and convenience stores. Others responded with variations on the same theme. Workers develop a keen sense of where they can access reproductive fundamentals, such as toilets, water and electrical outlets. Gary told me he keeps a mental list of the restaurants that allow couriers to use the toilet, and by extension, a list of those establishments 'that are dicks about it … They say, like, "oh, the restrooms are for customers only, sorry"'. Marcus told me that he scouts electrical outlets in public spaces:

> Sometimes I'll be sitting at a park and I look down – and I never would've noticed this before, but when they updated the app [for couriers] it just started, like, *draining* my battery. But yeah, sometimes I'll be sitting there, and I look down and see an outlet, like at the base of a lamp or on a concrete wall even. And you just walk over and plug in, no one cares … At first, like right when I first signed up, they [Caviar] gave you [an external] battery but, I don't know, maybe it got wet or something – it doesn't hold a charge anymore. So yeah, now I just keep an eye out … When I find a new [outlet], sometimes I'll just, like, real quick open up Google Maps and drop a pin so I remember.

For workers on Caviar and other delivery platforms, these public infrastructures – bike-share networks, water fountains, toilets and outlets – are reproductive resources, the means to stay powered and energised before performing the types of work that the platform recognises as labour.

REGULATORY AND RESEARCH IMPLICATIONS

Platforms are certainly not the first firms to exploit the accessibility and collective utility of urban infrastructures. Indeed, it is common for states or public agencies to invoke the economic benefits of infrastructure to justify investment (Leduc and Wilson 2013). It is also common for corporations to deploy some form of regulatory arbitrage and tax avoidance to minimise their overhead costs (Fleischer 2010), to benefit from the publicness of infrastructure while contributing nothing to its maintenance or upkeep. The platform economy is not unique in this regard.

Nonetheless, the question of urban infrastructure's role in the political economy of platform capitalism has been curiously under-appreciated. Two explanations account for this oversight. First, there is the tendency to disregard the infrastructure-like qualities of privately owned consumer spaces, such as supermarkets or convenience stores. Infrastructures, the thinking goes, are centrally coordinated, large-scale technical systems, usually owned, or at least managed or overseen, by public agencies. According to Richardson (2018), this view is a relatively recent invention. Platforms, she suggests, may be revitalising an older model of publicness, mediated less by the state than by market relations. Similarly to other platform urbanists (Barns 2018a, 2018b, 2019; Leszczynski 2020), Richardson strikes an optimistic, and anti-determinist tone, refusing to equate market-marketing with the deprivations of privatisation. However, again, this optimism may be overlooking the problems likely to arise as markets drive decision-making that affects the distribution of essential services and collective resources. There will be winners and losers: the winners will enjoy access to necessities, such as food, water, energy and information, while the losers' exclusion from infrastructures will perpetuate enduring urban inequalities and racialised segregations (Deener 2017; Graham and Marvin 2001).

However, there is also substance in Richardson's point. By amplifying the collective utility and accessibility of privately owned businesses such as supermarkets, we can begin to regulate them – as we do infrastructures and utilities – so that they must guarantee a more equitable geography of access. For platforms to play a productive role here, they need to come in from the regulatory cold; that is, to quit trying to shirk regulations designed to protect their workforce, to offer a just price for the labour and infrastructures upon which they depend in one way or another (Koehn and Wilbratte 2012).

The second explanation for infrastructure's absence from political economic critiques of platform capitalism is related to a great deal of the work required of platform labour being, by design, invisible. Sociologist David Hill (2020, p. 522) argues that the invisibility of platform operations – 'when platforms act to conceal their operation from the awareness of users' – creates overlapping 'injuries' that are at once cognitive, psychic and moral. The responsibility of a critical platform urbanism is to render the invisible visible, to counter those injuries by thinking beyond transactional data to its 'immanent outsides' (Massumi 2018, p. 9); that is, the reproductive work required to sustain platform labour as well as the value embedded in the collectively produced infrastructures of everyday urban life.

My hope for the concept of infrastructural surplus is that it helps us reinvigorate a politics centred on collective consumption goods. The two modes of infrastructural appropriation identified here – reformatting social space and transactional excess – suggest that we need both floors and ceilings as guardrails for market-mediated public infrastructures. The floor means investing in physical and social infrastructures. In the USA, when a single father drives part-time on Uber after his primary job, or when a college student works as an Instacart shopper, they are probably doing so to subsidise their depressed incomes against the rising costs of housing and education (for example, Hendrickson 2018). Investing in the publicness of infrastructures, from housing to education and healthcare, dampens the demand for precarious platform work. Markets may or may not be an adequate means for this investment; we need more and better tools for evaluating the regulatory failures of unfettered governance-by-market, and we need to remain vigilant to the evidence that markets are as liable to exploit vulnerabilities as they are to increase efficiency.

The ceiling means more aggressively bracketing collective goods and infrastructures from market predations to minimise the disparities and gaps that platforms open and exploit. Limits on the number of short-term rental units in cities where brokerage platforms, such as Airbnb, are exacerbating gentrification pressures, for example, or placing caps on the number of drivers on platforms, as Uber and Lyft do, are common regulatory responses. More must be done to ensure the publicness of scarce goods like housing and to expand the limited scope of the platform transaction to recognise the hidden, reproductive work involved in the job. A ceiling means reining in platforms' wildly inflated valuations by pegging their worth to the collective value they create. Why does Airbnb not fund affordable housing in gentrifying cities to buffer its role in gentrification? Why does Caviar not help finance safer biking infrastructure and protected lanes in the cities where it operates, especially after couriers have been killed making deliveries (Parry 2018)? Why did it take legislation to force Instacart to include hazard pay during the COVID-19 pandemic? These types of questions are not particular to technology firms. However, without regulation to ensure accessibility, collective use and worker well-being, platforms will never create value; they will only appropriate it.

Boosting the publicness of collective goods outside the platform and expanding the scope of the platform transaction to recognise the collective nature of those goods; these regulatory imperatives are two sides of the same coin. What good are public infrastructures if their value is co-opted through exploitative labour arrangements?

This chapter offered a preliminary sketch of infrastructural surplus as a contested source of value in platform urbanism. More empirical work is needed to understand the evolving arbitration of productive/visible and reproductive/ invisible work in platform labour, and how this arbitrage affects urban subjects and urban infrastructures differentially. For example, how do infrastructural appropriations refract across racial, classed and gendered identities in ways that manifest unevenly in platform labour (in)visibilities (Van Doorn 2017)? More work is also needed to identify the particularities of platforms' infrastructural exploits relative to companies outside the technology sector. What, if anything, is distinctive in the platform capture of infrastructural surplus – or is this just an old wolf in new sheep's clothing?

REFERENCES

Agre, P.E. (1994), 'Surveillance and capture: two models of privacy', *Information Society*, **10** (2), 101–27.
Amin, A. (2011), 'Re-thinking the urban social', *City*, **11** (1), 100–114.
Amin, A. (2014), 'Lively infrastructure', *Theory, Culture and Society*, **31** (7–8), 137–61.
Amin, A. and N. Thrift (2002), *Cities: Reimagining the Urban*, Cambridge: Polity.
Attoh, K., K. Wells and D. Cullen (2019), '"We're building their data": labor, alienation, and idiocy in the smart city', *Environment and Planning D: Society and Space*, preprint online, doi:10.1177/0263775819856626.
Barns, S. (2018a), 'Platform urbanism rejoinder: why now? What now?', *Mediapolis*, November, accessed 4 August 2020 at https://www.mediapolisjournal.com/2018/11/ platform-urbanism-why-now-what-now/.
Barns, S. (2018b), 'We are all platform urbanists now', *Mediapolis*, October, accessed 4 August 2020 at https://www.mediapolisjournal.com/2018/10/we-are-all-platform -urbanists-now/.
Barns, S. (2019), 'Negotiating the platform pivot: from participatory digital ecosystems to infrastructures of everyday life', *Geography Compass*, **13** (9), art. e12464.
Barry, E. (2016), 'Uber, Lyft effect on economy show work "innovation": case', podcast, 15 April, accessed 19 July 2020 at https://www.cnbc.com/2016/04/15/uber -lyft-effect-on-economy-show-work-innovation-case.html.
Bensinger, G. (2015), 'Rebuilding history's biggest dot-com bust', *Wall Street Journal*, 13 January, accessed 19 July 2020 at https://www.wsj.com/articles/rebuilding -historys-biggest-dot-come-bust-1421111794.

Birch, K. (2019), 'Technoscience rent: toward a theory of rentiership for technoscientific capitalism', *Science, Technology, and Human Values*, **45** (1), 3–33.

Bourdieu, P. (1977), *Outline of a Theory of Practice*, Cambridge: Cambridge University Press.

Bowles, R. (2020), 'Organized chaos: behind the scenes of Amazon's inventory management system', blog, accessed 19 July 2020 at https://www.logiwa.com/blog/amazon-inventory-management-system.

Brenner, N. (2013), 'Theses on urbanization', *Public Culture*, **25** (1), 85–114.

Briziarelli, M. (2018), 'Spatial politics in the digital realm: the logistics/precarity dialectics and Deliveroo's tertiary space struggles', *Cultural Studies*, **33** (5), 823–40.

Carse, A. (2012), 'Nature as infrastructure: making and managing the Panama Canal watershed', *Social Studies of Science*, **42** (4), 539–63.

Casilli, A. A. (2017), 'Digital labor studies go global: toward a digital decolonial turn', *International Journal of Communication*, **11**, 3934–54.

Castells, M. (1973), *The Urban Question: A Marxist Approach*, Cambridge, MA: MIT Press.

Daniels, A.K. (1987), 'Invisible work', *Social Problems*, **34** (5), 403–15.

Davidson, N.M. and J.J. Infranca (2015), 'The sharing economy as an urban phenomenon', *Yale Law and Policy Review*, **34** (2), 215–80.

De Stefano, V. (2015), 'The rise of the just-in-time workforce: on-demand work, crowdwork, and labor protection in the gig-economy', *Comparative Labor Law and Policy Journal*, **37** (3), 471–504.

Deener, A. (2017), 'The origins of the food desert: urban inequality as infrastructural exclusion', *Social Forces*, **95** (3), 1285–309.

DeVault, M.L. (2014), 'Mapping invisible work: conceptual tools for social justice projects', *Sociological Forum*, **29** (4), 775–90.

Etherington, D. (2013), 'Amazon's grocery business learns from Webvan that rapid growth is the enemy of fresh', *TechCrunch*, accessed 19 July 2020 at https://social .techcrunch.com/2013/06/17/amazons-grocery-business-learns-from-webvan-that -rapid-growth-is-the-enemy-of-fresh/.

Fast Company (2015), 'The world's 50 most innovative companies of 2015', accessed 19 July 2020 at https://www.fastcompany.com/most-innovative-companies/2015.

Fleischer, V. (2010), 'Regulatory arbitrage', *Texas Law Review*, **89** (2), 227–89.

Galloway, A.R. (2004), *Protocol: How Control Exists After Decentralization*, Cambridge, MA: MIT Press.

Gandy, M. (2005), 'Cyborg urbanization: complexity and monstrosity in the contemporary city', *International Journal of Urban and Regional Research*, **29** (1), 26–49.

Gorlick, S. (2020), 'In 2 years, Uber went from being in just a few cities to more than 100. How did Uber grow so quickly?', Twitter post, accessed 20 August 2020 at https://twitter.com/sgorlick/status/1295398477644664832.

Graham, S. and S. Marvin (2001), *Splintering Urbanism: Networked Infrastructures, Technological Mobilities and the Urban Condition*, New York: Routledge.

Greenberg, M. and P. Lewis (eds) (2017), *The City Is the Factory: New Solidarities and Spatial Strategies in an Urban Age*, Ithaca, NY: ILR Press.

Gregg, M. (2011), *Work's Intimacy*, London: Polity.

Guendelsberger, E. (2019), *On the Clock: What Low-Wage Work Did to Me and How It Drives America Insane*, New York: Little, Brown.

Hendrickson, C. (2018), 'The gig economy's great delusion', *Boston Review* political forum, 8 January, accessed 20 July 2020 at http://bostonreview.net/class-inequality/clara-hendrickson-gig-economys-great-delusion.

Herman, R. and J.H. Ausubel (1988), 'Cities and infrastructure: synthesis and perspectives', in J.H. Ausubel and R. Herman (eds), *Cities and Their Vital Systems: Infrastructure Past, Present, and Future*, Washington, DC: National Academies Press, pp. 1–21.

Hill, D.W. (2020), 'The injuries of platform logistics', *Media, Culture and Society*, **42** (4), 521–36.

Huws, U. (2016), 'Logged labour: a new paradigm of work organisation?', *Work Organisation, Labour and Globalisation*, **10** (1), 7–26.

Instacart (2019), 'Terms of service', accessed 19 July 2020 at https://www.instacart .com/terms.

Jackson, S.J. (2014), 'Rethinking repair', in T. Gillespie, P.J. Boczkowski and K.A. Foot (eds), *Media Technologies: Essays on Communication, Materiality, and Society*, Cambridge, MA: MIT Press, pp. 221–40.

Jarrahi, M.H. and W. Sutherland (2019), 'Algorithmic management and algorithmic competencies: understanding and appropriating algorithms in gig work', in N.G. Taylor, C. Christian-Lamb, M.H. Martin and B. Nardi (eds), *Information in Contemporary Society*, Cham: Springer International, pp. 578–89.

Kanngieser, A. (2013), 'Tracking and tracing: geographies of logistical governance and labouring bodies', *Environment and Planning: Society and Space*, **31** (4), 594–610.

Koehn, D. and B. Wilbratte (2012), 'A defense of a Thomistic concept of the just price', *Business Ethics Quarterly*, **22** (3), 501–26.

Kolodny, L. (2016), 'Instacart's app has changed grocery stores for good', *TechCrunch*, 31 August, accessed 19 July 2020 at https://social.techcrunch.com/2016/08/31/ instacarts-app-has-changed-grocery-stores-for-good/.

Langley, P. and A. Leyshon (2017), 'Platform capitalism: the intermediation and capitalisation of digital economic circulation', *Finance and Society*, **3** (1), 11–31.

Leberstein, S. (2016), 'Uber's car leasing program turns its drivers into modern-day sharecroppers', accessed 20 August 2020 at https://qz.com/700473/ubers-car-leasing -program-turns-its-drivers-into-modern-day-sharecroppers/.

Leduc, S. and D. Wilson (2013), 'Roads to prosperity or bridges to nowhere? Theory and evidence on the impact of public infrastructure investment', *NBER Macroeconomics Annual*, **27** (1), 89–142.

Leszczynski, A. (2020), 'Glitchy vignettes of platform urbanism', *Environment and Planning D: Society and Space*, **38** (2), 189–208.

Lianos, M. and M. Douglas (2000), 'Dangerization and the end of deviance: the institutional environment', *British Journal of Criminology*, **40** (2), 261–78.

Mackenzie, A. (2005), 'Untangling the unwired: Wi-Fi and the cultural inversion of infrastructure', *Space and Culture*, **8** (3), 269–85.

Manjoo, F. (2014), 'Grocery deliveries in sharing economy', *New York Times*, 21 May, accessed 6 August 2020 at https://www.nytimes.com/2014/05/22/technology/ personaltech/online-grocery-start-up-takes-page-from-sharing-services.html.

Manjoo, F. (2015), 'Instacart's bet on online grocery shopping', *New York Times*, 29 April, accessed 21 August 2020 at https://www.nytimes.com/2015/04/30/ technology/personaltech/instacarts-bet-on-online-grocery-shopping.html.

Mascarenhas, N. (2020), 'Instacart announces new COVID-19 policies and plans to hire 250,000 more shoppers', *TechCrunch*, accessed 21 August 2020 at https://social .techcrunch.com/2020/04/23/instacart-announces-new-covid-19-policies-and-plans -to-hire-250000-more-shoppers/.

Massumi, B. (2018), *99 Theses on the Revaluation of Value: A Postcapitalist Manifesto*, Minneapolis, MN: University of Minnesota Press.

Mathew, B. (2008), *Taxi! Cabs and Capitalism in New York City*, Ithaca, NY: Cornell University Press.

Mattern, S. (2018), 'Maintenance and care', *Places Journal*, November, doi:10.22269/181120.

McCreight, G. (2019), 'The story behind an Instacart order, part 1: building a digital catalog', accessed 21 August 2020 at https://tech.instacart.com/the-story-behind-an-instacart-order-part-1-building-a-digital-catalog-46df5a8ff705.

Mezzadra, S. and B. Neilson (2017), 'On the multiple frontiers of extraction: excavating contemporary capitalism', *Cultural Studies*, **31** (2–3), 185–204.

Milbourn, T. (2015), 'In the future, employees won't exist', *TechCrunch*, accessed 8 August 2020 at https://social.techcrunch.com/2015/06/13/in-the-future-employees-wont-exist/.

Miller, J. (2014), 'The fourth screen: mediatization and the smartphone', *Mobile Media and Communication*, **2** (2), 209–26.

Nickelsburg, M. (2020), 'Instacart threatens to pull out of Seattle if new hazard pay law passes', *GeekWire*, accessed 19 August 2020 at https://www.geekwire.com/2020/instacart-threatens-pull-seattle-new-hazard-pay-law-passes/.

Parry, T.F. (2018), 'The death of a gig worker', *The Atlantic*, 1 June, accessed 23 December 2019 at https://www.theatlantic.com/technology/archive/2018/06/gig-economy-death/561302/.

Prassl, J. (2018), *Humans as a Service: The Promise and Perils of Work in the Gig Economy*, Oxford: Oxford University Press.

Richardson, L. (2018), 'Platforms and the publicness of urban markets', *Mediapolis*, October, accessed 5 August 2020 at https://www.mediapolisjournal.com/2018/10/platforms-and-the-publicness-of-urban-markets/.

Richardson, L. (2020), 'Platforms, markets, and contingent calculation: the flexible arrangement of the delivered meal', *Antipode*, **52** (3), 619–36.

Rodgers, S. and S. Moore (2018), 'Platform urbanism: an introduction', *Mediapolis*, October, accessed 19 July 2020 at https://www.mediapolisjournal.com/2018/10/platform-urbanism-an-introduction/.

Rosenblat, A. (2018), *Uberland*, Oakland, CA: University of California Press.

Rossi, U. (2019), 'The common-seekers: capturing and reclaiming value in the platform metropolis', *Environment and Planning C: Politics and Space*, **37** (8), 1418–33.

Rossiter, N. (2016), *Software, Infrastructure, Labor: A Media Theory of Logistical Nightmares*, New York: Routledge.

Sadowski, J. (2020a), 'Cyberspace and cityscapes: on the emergence of platform urbanism', *Urban Geography*, **41** (3), 448–52.

Sadowski, J. (2020b), 'The Internet of landlords: digital platforms and new mechanisms of rentier capitalism', *Antipode*, **52** (2), 562–80.

Schaaf, M. (2020), 'Building an essential service during a pandemic', accessed 21 August 2020 at https://tech.instacart.com/building-an-essential-service-during-a-pandemic-3e2e51616a45.

Shapiro, A. (2018), 'Between autonomy and control: strategies of arbitrage in the "on-demand" economy', *New Media and Society*, **20** (8), 2954–71.

Srnicek, N. (2016), *Platform Capitalism*, Malden, MA: Polity.

Star, S.L. and A. Strauss (1999), 'Layers of silence, arenas of voice: the ecology of visible and invisible work', *Computer Supported Cooperative Work (CSCW)*, **8** (1–2), 9–30.

Stehlin, J.G. (2019), *Cyclescapes of the Unequal City: Bicycle Infrastructure and Uneven Development*, Minneapolis, MN: University of Minnesota Press.

Sutherland, W., M.H. Jarrahi, M. Dunn and S.B. Nelson (2020), 'Work precarity and gig literacies in online freelancing', *Work, Employment and Society*, **34** (3), 457–75.

Teubner, G. (2020), 'The constitution of non-monetary surplus values', *Verfassungsblog*, 3 March, accessed 19 July 2020 at https://verfassungsblog.de/the-constitution-of -non-monetary-surplus-values/.

The Economist (2014), 'There's an app for that', *The Economist*, 31 December, accessed 23 May 2017 at http://www.economist.com/news/briefing/21637355 -freelance-workers-available-moments-notice-will-reshape-nature-companies-and.

Van Dijck, J. (2014), 'Datafication, dataism and dataveillance: big data between scientific paradigm and ideology', *Surveillance and Society*, **12** (2), 197–208.

Van Doorn, N. (2017), 'Platform labor: on the gendered and racialized exploitation of low-income service work in the "on-demand" economy', *Information, Communication & Society*, **20** (6), 898–914.

Wahl, C. (in progress), *Racing in the Aisles: How Algorithmic Management Structures Labor in On-Demand Grocery Work*, Philadelphia, PA: University of Pennsylvania.

Weeks, K. (2007), 'Life within and against work: affective labor, feminist critique, and post-Fordist politics', *Ephemera: Theory and Politics in Organization*, **7** (1), 233–49.

Wells, K.J., K. Attoh and D. Cullen (2020), '"Just-in-place" labor: driver organizing in the Uber workplace', *Environment and Planning A: Economy and Space*, 0308518X20949266.

Whitney, A. (2017), 'What it's really like to be an Instacart shopper', *bon appétit*, 26 September, accessed 19 July 2020 at https://www.bonappetit.com/story/instacart -shopper-for-a-day.

6. Dual value production as key to the gig economy puzzle[1]

Niels van Doorn and Adam Badger

THE GIG ECONOMY PUZZLE: THIN MARGINS AND OUTSIZED EXPECTATIONS

The puzzle we attempt to solve in this chapter is how gig economy companies can continue to grow their business despite regularly incurring huge losses.[2] While the first step towards solving this puzzle is easily made, by bringing into focus the crucial role of venture capital and investment firms, this immediately requires us to confront a more puzzling reality: that these firms have continued to fund loss-making gig companies operating in industries with extremely thin margins. To make sense of investors' high expectations, we believe it is necessary to start by asking a deceptively basic question: what type of work is platform-mediated gig work? Phrased differently: what types of value are created through platform labour?

To answer this question, it may be strategically useful to momentarily accept the position defended by gig economy companies in various court cases, namely, that they merely provide the technical platform on which service providers find access to their customer base (for example, Tomassetti 2016). From this perspective, these companies provide an informational service that is categorically distinct from the service provided by the gig worker and therefore they should not – indeed cannot – be legally held accountable as employers (for discussion of this, see Aloisi 2015; Meijerink and Keegan 2019; Stewart and Stanford 2017). In return for this service, the argument continues, gig

---

[1] This chapter is a partly reworked version of the following article previously published in Antipode: Van Doorn, N. & Badger, A. (2020), Platform capitalism's hidden abode: producing data assets in the gig economy. *Antipode*, 52(5), 1475–1495.

[2] See for example, Hawkins (2019) reporting Uber's $5.4billion loss in a single quarter, or Butler (2019) reporting Deliveroo losses of £232million in the same period. Neither company has ever turned a profit.

economy companies charge a commission on each transaction conducted via their platform.

Crucially, however, in addition to extracting rent from each transaction they orchestrate, platforms also extract data about these transactions, and usually about a lot more, which means that gig workers can likewise be understood to provide an informational service to the platforms they use. That this service is neither optional nor remunerated suggests that the data extraction 'continues to open up new frontiers for the expansion of the logics of property and to blur the borders between processes of governance and dynamics of capitalist valorisation' (Mezzadra and Neilson 2017, p. 195). That is, gig work is a form of data work and the gig economy should be understood as one salient phenomenon within the broader framework of financialised platform capitalism (Langley and Leyshon 2017; Srnicek 2017).

In our view, the digital platform is one of capital's new frontiers in its ongoing expansionist drive amid decreasing interest rates, allowing it to expand into previously uncharted areas of life through data- and finance-driven modes of accumulation. Platform capitalism forms a (provisional) solution to the problem of capital appreciation facing the investor class. Yet one class's solution is another class's problem, given that an economy governed by platforms is ultimately realising a massive redistribution of wealth and risk; the former from the working class to the investor class and the latter from employers to workers. Moreover, what exacerbates this problem is that not enough is done to counter it. There is a serious lack of robust and effective regulation in respect of the gig economy, which is largely the result of a chronic lack of expertise and political will among public officials. While we cannot solve the problem of political will, scholars can assist policy-makers as regards building expertise, and it is with this goal in mind that this chapter has been written.

To begin solving our puzzle, then, we introduce the notion of 'dual value production', which describes how platforms capture two types of value from gig work: the monetary value associated with the service transaction, and the more speculative and volatile types of value associated with the data generated during service provision. We then elaborate on the construction of data as a specific asset class and consider the process and consequences of data assetisation. Shifting our perspective from the platform to gig workers, we subsequently discuss two grassroots initiatives that resist the unbridled data extraction from gig work and attempt to reclaim their data assets. The next section takes another step towards solving our puzzle, as we move up the value chain and examine the role of what we term 'meta-platforms'. It is on this level that we are confronted with the true power brokers of the platform economy – as well as with the magnitude of the puzzle at hand – and we therefore end our chapter by proposing an ambitious set of regulatory and policy measures that could curb this unprecedented power. First, however, we offer a brief discussion of how our study positions itself vis-à-vis existing gig economy research, what research puzzle – that is, knowledge gap – we aim to solve, and what methods we have deployed to accomplish this.

THE RESEARCH PUZZLE

Our approach to platform-mediated gig work deviates from, while remaining indebted to, what we take to be the two main strands of gig economy research that have so far shaped this field: (1) labour process theory-inspired scholarship concerned with algorithmic management and information asymmetries (for example, Gandini 2019; Rosenblat and Stark 2016; Veen et al. 2020) and (2) legal scholarship primarily focused on the social costs of worker misclassification (for example, Aloisi 2015; Prassl 2018). Both research strands share a similar analytical scope, in so far as associated studies critically attend to how the precarious conditions and misclassification of gig workers are enforced through technological and legal means. That is, gig economy research, including our own contribution, has so far mostly restricted itself to the sphere of the platform as both a business model and work environment. In contrast, here we aim to expand this purview in order to examine the broader political economy of data and finance capital that not only keeps gig platforms open for business, but also demands increasingly strict discipline over how such business is conducted, resulting in progressively worse working conditions and decreasing wages.

The specific research puzzle we aim to solve in this chapter thus pertains to a knowledge gap left by current gig economy scholarship: what happens in the space between the intensifying exploitation of gig workers and the massive market valuation of non-profitable gig economy companies? To solve this puzzle, it is necessary to take a multidisciplinary approach that draws on insights from political economy, platform studies, critical data studies and the (digital) sociology of work. Our analytical perspective follows the contours of our respective research projects, which both examine platform-mediated labour. We have each spent substantial periods conducting (auto-)ethnographic research, during which we not only studied gig workers, but also engaged in gig work ourselves. Van Doorn spent two years studying application (app)-based food delivery and domestic cleaning services in New York, Berlin and Amsterdam (spending eight months in each city), also working as a courier and cleaner in the latter two cities.[3] Badger has similarly undertaken

---

[3]   This fieldwork consisted of participant observation on city streets, in homes and offices, and in online spaces. It also resulted in 158 formal semi-structured interviews, primarily with couriers and cleaners, but also with some entrepreneurs in food delivery,

food delivery work for two platforms in London over a period of nine months, in addition to carrying out 18 months of ethnographic research within a grass-roots trade union responsible for organising gig workers.[4]

In addition to our long-term ethnographic studies of gig workers' every-day experiences, we have also conducted extensive desk research on the institutional, financial/economic and technological conditions enabling the platformisation of low-wage service work across local and national settings. These analyses extended beyond the Global-North purview of our respective ethnographies and allowed us to identify similar dynamics and developments in other parts of the world. Our shared interest in the political economy of low-wage gig work, crystallised in app-based food delivery, is what brought us together and pushed us to jointly examine in more detail the role that data plays in the daily operations and business models of gig platforms. We focus on data extraction in low-wage gig work as this is a phenomenon that spans several quickly growing global industries, where it serves to increase the rate of exploitation of often vulnerable, migrant workers who have no say over how their data is used and valorised.

THE KEY TO THE PUZZLE: DUAL VALUE PRODUCTION

In this chapter we argue that gig work under conditions of platform capitalism is characterised by a process that we term 'dual value production': the mone-tary value produced by the service provided is augmented by the use and spec-ulative value of the data produced before, during and after service provision.[5] Platforms capture part of this monetary value by charging rent, in the form of a commission, while capturing all of the value produced by gig workers' data

---

cleaning and adjacent industries in the three cities. Many more informal conversations took place during this two-year period.

[4]   This fieldwork consisted of participant observation across a range of digital and urban spaces. Beyond regular informal workplace conversations, 14 formal semi-structured interviews were conducted. Beyond participant observation and inter-view methods, data collection included a mix of video, photography and audio record-ings, creating multi-media diary entries that express the complexity of the workplace.

[5]   What distinguishes the business model of labour platforms from that of other platform companies, such as Facebook or Google, is that the latter's revenues are pri-marily dependent on the advertising-driven ecosystems they create and manage. In con-trast, the revenues of labour platforms depend primarily on worker exploitation and rent-seeking (cf. Srnicek 2017). While data capture is crucial to both platform business models, dual value production is a unique characteristic of labour platforms because gig workers produce data while providing paid services via the platform, whereas users of Google or Facebook produce data as consumers of these platforms.

labour. That is, using Sadowski's (2019, p. 10) pithy formulation, 'platforms collect monetary rent and data rent'. Yet, whereas the value of this monetary rent can be dynamically determined by the platform, the value of data rent is fundamentally indeterminate in so far as it derives from speculative and performative practices.

Platforms engage in constant data accumulation because of the potential value this data, once processed by their analytics software, might embody or give rise to.[6] This value derives, in part, from data's expected or actual practical utility in operational processes (that is, achieving functional goals and systems optimisation). Yet captured data also attracts venture capital and grows financial valuations, to the extent that investors expect data-rich platform companies to achieve competitive advantages by creating data-driven cost efficiencies, cross-industry synergies and new markets. In this way, it becomes possible 'to convert data into money' (Sadowski 2019, p. 11), which is then again invested in activities and technologies that increase the capture of data.

While data may at first seem like a supplementary component of the service provided, it is in reality key to understanding what gig platforms are about. Focusing on datafication allows us to grasp how app-governed gig workers function as pivotal conduits in software systems that combine distributed data generation and centralised analytics, depending on layers of existing (public and private) urban infrastructure, from free Wi-Fi networks to roads and bike lanes (Shapiro 2017). In practice, a courier's smartphone and physical labour become a site of translation through which complex urban environments are formatted into machine-readable data streams. These apparatuses thereby produce digital data as a particular asset class (Sadowski 2019, 2020), which is central to platform capitalism 'as a mode of accumulation that is simultaneously a system of domination' (Fraser 2016, pp. 164–5). In the next section we explain what this means.

---

[6]   As Sadowski writes (2019, p. 10): 'the value of data is uncertain; the valuation of data is complex'. Moreover, as he notes elsewhere, '[t]he conditions needed to convert data capital into economic capital may never arrive, but that does not stop the cycle of accumulation' (Sadowski 2020, p. 572). This is because the costs associated with the capture, storage and processing of data are relatively insignificant for platform companies flush with investment capital, especially compared with the funds required for marketing (for example, advertising, discounts and promotions), lobbying and litigation.

DATA ASSETISATION: A PREDATORY PLATFORM SOLUTION

What type of asset is captured data? A main distinguishing feature of the data asset is its high-value elasticity, that is, both its operational use value and its speculative financial value tend to increase significantly as it scales. To elucidate this elasticity, it is helpful to return to our notion of dual value production on food delivery platforms. On the level of service provision, a platform company's bottom line (that is, net income) consists of the rent the platform extracts from each completed food order (that is, the commission it takes from the restaurant plus the delivery fee it charges the customer, together forming its top-line revenue) minus the piece-rate labour costs associated with each order and other expenses, such as price discounts for consumers to grow market share. In traditional Marxist analyses, this is the scene of exploitation: 'recompensed only for the socially necessary cost of their own reproduction, [food delivery workers] have no claim on the surplus value their labour generates, which accrues instead to the [platform company]' (Fraser 2016, p. 164). However, as Fraser argues, the problem with this perspective is that, by focusing on 'capital's exploitation of wage labour in commodity production' (Fraser 2016, p. 165), it marginalises another fundamental process that is both entangled with exploitation and operates as its racialised condition of possibility: expropriation, or what David Harvey (2005) has termed accumulation by dispossession.

Expropriation 'works by *confiscating* capacities and resources and *conscripting* them into capital's circuits of self-expansion' (Fraser 2016, p. 166, original emphases), which accurately describes the globe-spanning capture of data assets produced by mostly (im)migrant food delivery workers who lack ownership or meaningful control over these assets (cf. Couldry and Mejias 2019). Moreover, data expropriation makes it possible for food delivery platforms to continually optimise their accumulation strategies based on exploitation, for instance, by dynamically adjusting – while progressively decreasing – riders' delivery fees based on aggregated market data in order to increase profit margins (Van Doorn and Chen forthcoming). Data expropriation is a practice characterised by alienation and unfreedom, which forms the condition of possibility for the exploitation of food delivery workers who, as independent contractors, are nominally free to choose when or for how long they work and which orders they accept. It is precisely these sequences of decision-making activities from which data assets can be derived, which means that couriers' freedom of choice can be strategically leveraged as a behavioural informational service that can be used against their best interests.

However, whereas the unit economics of courier exploitation expands in a linear fashion, the captured data assets expropriated from each courier only become actionable once their accumulation reaches scale, after which their value grows exponentially.[7] This, then, is what it means to say that captured data is a highly elastic asset class: the value associated with its expropriation is much more sensitive to the qualities of scale than is exploitation of labour. We should be careful here, however, not to naturalise the notion of scale and to avoid conflating it with volume or size. It would be more accurate to say that scale is an effect of a platform company's data analytics capacities. That is, ownership and control over the computational architecture built for data capture is essential.

This is illustrated most clearly in the initial public offering (IPO) filings of established gig-economy companies. For example, Uber Technology, Inc.'s (2019, pp. 155–6) filing states:

> Managing the complexity of our massive network and harnessing the data from over 10 billion trips exceeds human capability, so we use machine learning and artificial intelligence, trained on historical transactions, to help automate marketplace decisions. We have built a machine learning software platform that powers hundreds of models behind our data-driven services across our offerings and in customer service and safety.

In this constellation, data capturing sensors, machine learning algorithms and gig workers do not function in isolation. Instead, they form vital interlocking components that converge into one system and allow it to (operate at) scale.

This positive feedback loop, between a data-producing labour process and algorithmic systems that self-optimise as they analyse this data, is at the heart of machine learning's promise of full automation. Importantly, this promise drives the operational practices and investor pitch decks of food delivery startups and other gig economy companies worldwide. In their shared vision, one of the key value propositions of digital platforms is that their data analytics capacity will eventually enable the automation of all fungible forms of gig work, thus diminishing contracted labour costs to zero. The platform company to first accomplish this goal will subsequently conquer the market and reach

---

[7]   Reaching scale is a constantly moving target. It is a dynamic site of experimentation that is contingent not just on a company's evolving operational goals, but also on environmental variables such as the nature, prevalence or relative significance of the activity being captured and datafied. Data analytics is not an exact science and, although the accumulation of more data generally increases the accuracy and versatility of predictive calculations, it is impossible to predict in advance at which threshold an expanding data set, or combination of data sets, will increase in value by becoming more actionable.

monopolistic status. Gig workers, within this speculative vision, will have (unwittingly) contributed to their own inevitable obsolescence.

SOLUTIONS FROM BELOW: GIG WORKERS RECLAIMING THE DATA ASSET

Not so fast. Gig workers are not the hapless exploitable dupes that Silicon Valley too easily takes them for, and their data assets have recently emerged as a new frontier for organised resistance. Fed up with decreasing wages, degrading working conditions and persistent information asymmetries, workers are seeking new ways to access, own and leverage their data in order to win back power in the gig economy. Here we discuss two important and inspiring grassroots initiatives, addressing both their potential and their limitations.

The Worker Info Exchange: Weaponising the General Data Protection Regulation

The Worker Info Exchange (WIEx[8]) is an initiative led by former Uber driver James Farrar, who is also a lead claimant in an ongoing UK court case against the company's alleged misclassification of its workforce. The WIEx brings together workers, academics, lawyers and computer scientists in an effort to not only gain legal access to driver-generated data, but also to build a computational infrastructure capable of mining this data for useful insights. To accomplish this, Uber drivers are encouraged to submit subject access requests (SARs) and then contribute the data they receive to a larger data pool collected, managed and analysed by the WIEx. Although the computational power available will remain vastly limited in comparison to Uber's capture apparatus, the hope is that the collected data will nevertheless reveal information on topics such as payment, management of driver supply, worked hours and the company's use of reputational data. This information could then be utilised in court to establish inconsistencies and falsehoods in Uber's claims, thereby poking holes in its legal defence and challenging the contractual arrangement that currently deprives drivers of (collective) power (Holder 2019). The key move here is not just the computational analysis of SAR-acquired data sets, but the organised collection of these data sets at a scale previously unseen.

The WIEx's efforts, while enabled by Europe's General Data Protection Regulation (GDPR), are inspired, in part, by New York City's pioneering new ride-hail legislation, whose licence cap and accompanying minimum wage

---

[8]   See https://workerinfoexchange.org/ for more information (accessed 23 April 2021).

regulations could not have been accomplished without the city government's sustained pressure on Uber and Lyft to turn over detailed trip data (Holder 2019). In both cases, data is understood as integral to challenging corporate platform power and ensuring public welfare, especially the welfare of precarious ride-hail drivers who have seen their wages decrease over the past few years. However, a critical limitation of these efforts is their reliance on access and transparency, both of which are dependent on private companies' readiness to accommodate requests and their compliance with existing regulatory frameworks, something these companies do not have a great track record on. Meanwhile, the fundamental power imbalance that subtends (platform) capitalism, predicated on asset ownership, is left intact.

Coopcycle: From Data Access to Ownership

The issue of ownership brings us to our second example. Coopcycle, based in France, describes itself as 'the European federation of bike delivery coops. Governed democratically by coops, it enables [these coops] to stand united and to reduce their costs thanks to resources pooling. It creates a strong bargaining power to protect the bikers [*sic*] rights'.[9] The resources it pools include services such as a software platform (distributed under Coopcycle's custom-made CoopyLeft licence, which prohibits use by non-cooperative businesses), a delivery app, administrative and legal support, and shared drafting of funding proposals. Whereas the WIEx focuses on the piecemeal collection of driver data made accessible by GDPR legislation, Coopcycle moves several steps ahead, taking the production, analysis and monetisation of delivery data into its own hands by building a collectivised computational architecture that could grant durability and scale to associated bike delivery cooperatives.[10] This endeavour begins to address a problem that has so far hindered the success of individual platform cooperatives, namely, their struggle to compete with the scaling capacities and seemingly unlimited resources of corporate platforms (Van Doorn, 2017).[11]

Sidestepping conditional and/or limited access to private data assets, Coopcycle prioritises collective ownership of data assets as a means to achieve worker power and autonomy. These commonly owned data assets can be leveraged in various operational and commercial activities, from the optimisation

---

[9]   See https://coopcycle.org/en/ for more information (accessed 23 April 2021).
[10]   For a more comprehensive discussion of platform cooperatives, see Chapter 9 in this volume.
[11]   Coopcycle shares this function and mission with the Platform Cooperativism Consortium, an international 'hub that starts, grows, and converts platform coops' (see https://platform.coop/who-we-are/pcc/, accessed 23 April 2021).

of the delivery process to the negotiation of transactions with clients, funders and other third parties. Currently, however, the key challenge is market penetration, given that one large group of potential clients – restaurants – remains tied to application programming interface (API) connections and service contracts with corporate delivery platforms. This is further hindered by the continual update cycle of mobile operating systems (that is, iOS and Android). That is, each time these operating systems are updated, the apps that rely on them need to be updated accordingly. While this is a relatively simple task for companies with large technology teams, such as Deliveroo and Uber, it poses a more substantial challenge when sustained access to technical resources is limited. The consequent impact on user experience, combined with a minimal operating budget, are likely to have a negative impact on user retention and growth. Without a growing portfolio of clients that can be served on a daily basis, data assets cannot be accumulated or exploited in the first place.

THE PUZZLE EXPANDS: FROM PLATFORM TO META-PLATFORM

Ultimately, as valuable as Coopcycle's efforts to socialise bike delivery coops undoubtedly are, its development of worker-owned economies of scale and collaborative software synergies pales in comparison to the type of massively bankrolled synergetic activities taking place elsewhere, at a scale that exceeds the purview of any individual platform company. To solve our gig economy puzzle and grasp what keeps deeply unprofitable platform companies afloat, we have to move up one tier in the rent-seeking value chain of financialised platform capitalism. This tier is the domain of what we term 'meta-platforms'; venture capital firms and investment funds looking to exploit the network effects and synergetic possibilities that emerge when managing a large and varied portfolio of investments in platform companies and other data-centric businesses, each intent on disrupting different industries by leveraging their analytics capacities.

We use the term 'meta-platform' because the growing power of these financial institutions stems from how they effectively operate as higher-order platforms whose profits are constituted by the rents extracted every time it matches investors, including institutional investors, such as pension funds and sovereign wealth funds, with technology companies looking for capital injections that will allow them to continue to scale quickly.[12] Paying critical

---

[12]   For instance, during Lyft's recent IPO roadshow, the company repeated its assertion that prioritising data-driven growth and innovation over short-term profits is the right business strategy (Lyft, Inc. 2019). As long as potential investors can be

attention to meta-platforms also moves us beyond a narrow concern with shareholder value, in so far as the stakes of our analysis do not just pertain to the influence of shareholder objectives on a company's daily operations, but demand that we account for the strategic governance of mutually reinforcing monopoly formations across sectors.

The meta-platform par excellence is SoftBank, the conglomerate that manages the $100 billion Vision Fund, nearly half of which is financed by Saudi Arabia's sovereign wealth fund. According to SoftBank's founder and Chief Executive Officer, Masayoshi Son, Vision Fund's portfolio companies control 90 per cent of the ride-hailing market worldwide (Alpeyev 2019), which is a percentage that should give us pause for thought. Son's approach, especially since the inauguration of the Vision Fund, has been to over-invest in particular platform companies and thereby aim to pre-ordain a winner in various competitive markets. This then sets up Son's 'cluster of number ones' strategy, which revolves around the creation of productive synergies between portfolio companies 'whose whole is theoretically greater than the sum of its parts – an added value derived from the partnerships and business opportunities that come with being a part of the SoftBank family' (Medeiros 2019). These partnerships and business opportunities largely focus on finding ways to actualise the potential of immense amounts of data captured from a great variety of sources. A recent *Wired* article summarises Son's vision as:

> a future where every time that we use our smartphone, or call a taxi, or order a meal, or stay in a hotel, or make a payment, or receive medical treatment, we will be doing so in a data transaction with a company that belongs to the SoftBank family. And, as Son likes to say: 'Whoever controls data controls the world.' (Medeiros 2019, n.p.)

Meta-platforms seek to control the world, or at least the platform ecosystems that increasingly reshape the world in their image. Having learned expensive lessons in the wake of the dot.com collapse, during which Son suffered a stunning $70 billion loss (Sherman 2019), meta-platform executives now aim to construct data-centric architectures of durability that will protect them in case the next technology bubble bursts – a bubble that they themselves will have helped to create. Even in the event that Uber folds (for instance, because governments around the world finally agree that the company is an employer and investors would consequently lose interest in its shares), its IPO has offered SoftBank an opportunity to cash out some of its equity and use these returns to invest in, and thereby anoint, the next Uber.

---

convinced that a platform company could at one point attain monopoly-like status, it can expect new capital injections that subsidise its ongoing efforts to gain market share and improve its financial performance.

It seems likely, however, that SoftBank would abstain from further investments in risky gig economy companies, instead opting to invest in the next Palantir (Peter Thiel's data-mining firm), or a startup that would complement its current investee, Arm (a British semiconductor and software design company that has become a major player in artificial intelligence, AI, development). While SoftBank's shock-and-awe investment strategy has generated both frustration and marvel among investors and analysts, its recent mishandling of the WeWork debacle, which resulted in the cancellation of the firm's IPO, painfully illustrated the fallibility of its model or vision (Alpeyev et al. 2019). Since then, SoftBank and its Vision Fund have been under increased financial pressure and scrutiny, as the firm seeks to stay afloat by selling up to $41 billion of assets at a discount in order to buy back its shares (Nussey 2020). In this shift 'from long-term domination to short-term survival' (Sherman 2019), SoftBank demonstrates its fealty to shareholders at the expense of its startup portfolio, as platform companies are increasingly expected to show a road to profitability by cutting costs, laying off employees and selling off operating units (Ongweso 2020).

SOLUTIONS FROM ABOVE: POSSIBLE PUBLIC RESPONSES

While platforms come and go, meta-platforms allocating the wealth of nations are becoming too big to fail. It is this massive privatisation of public wealth that returns us to the position and plight of gig workers under conditions of financialised platform capitalism. While it is true that finance capital subsidises a large share of gig workers' daily wages, it is equally true that it ultimately seeks to render their labour obsolete. Meanwhile, its investment comes with stipulated expectations and constraints in respect of how a platform company can run its business, pushing a high-risk/high-gain model that has valued rapid growth and limited liability. In times of crisis, as this model becomes destabilised, we see how platforms that cannot weather the strain become expendable in a manner that mimics the disposability of gig workers, just further upstream.

These dynamics demonstrate the need for regulatory measures that likewise have a forceful upstream impact. In order to organise a concerted pushback against the massive power of meta-platforms, we need regulatory intensification as well as policy innovations that hit platform capitalism's investor class where it hurts. To pave the way for such interventions, policy-makers and academics alike should develop coordinated research projects that take both a top-down and bottom-up approach. By bringing into productive conversation political economy research, platform studies and ethnographic fieldwork, the different operative scales (and their interrelated puzzles), explored above, become more knowable and therefore actionable. That is, we should simultaneously follow the money and the worker if we are to fully grasp and regulate the gig economy. Ultimately, we think, this should result in the abolition of the gig economy as we know it, so that from its ashes may rise an economy built on solidarity instead of exploitation and expropriation. To conclude this chapter, we briefly suggest some proposals that should take us in this direction, moving from forms of regulatory intensification focused on existing (meta-) platforms, toward public policies and investments that could foster new platform-based initiatives.

Regulatory Intensification

The first, most straightforward progressive move is to more strictly regulate existing platform companies and to enforce this regulation across the board. Any rule or law is only as good as its enforcement and, owing to a structural lack of institutional capacities and political will, gig economy platforms have for too long been able to determine the rules of the game. While gig worker reclassification and its enforcement will be one part of the solution, it will not be an adequate measure if divorced from a broader set of regulations that seek to curb the widespread commodification of low-wage labour across industries (Van Doorn et al. 2020). Improving wages, working conditions, and social protections for all workers, regardless of employment or residency status, will create a redefined and more equitable playing field in which workers (particularly migrants and minorities) have access to better jobs and will no longer have to resort to platform-mediated gigs.

In addition to labour regulation, stricter tax legislation is also a crucial weapon in the public arsenal. Here we should not only think of higher corporate taxes for platform companies in general terms, but more specifically consider frameworks for international coordination that aim to close global tax loopholes and end the rampant regulatory arbitrage that companies such as Uber engage in (Browning and Newcomer 2019). Countries such as the Netherlands, Ireland and Singapore should be forced to eliminate their tax havens and stop luring technology companies with tax breaks and other forms of corporate welfare. Moreover, instead of rewarding platform companies for their losses, by tying corporate income taxation to profits, this taxation should instead be based on a company's revenues. Another strategy is to create a special tax on big data-generated revenues (Madsbjerg 2017), although it is notoriously difficult to assign monetary value to data and it would likewise be challenging to ascertain how much data gig platforms sell to third parties or otherwise leverage toward revenue generation.

The topic of data leads us to another area where more forceful regulation is needed, namely, data rights pertaining to access, control and ownership of data by platform workers as well as other end users. We have already seen that initiatives such as the WIEx are pushing this agenda among gig workers, but much more could be done to support these efforts. While the GDPR offers a transnational framework for pursuing the data rights of gig workers, its focus on personal data posits severe limitations on its applicability and thus a more comprehensive and synthetic approach is required that leverages the most useful elements from various other legal frameworks currently operative within the European Union (Gallagher et al. 2019). Beyond data rights, we should also regulate for increased transparency and oversight of platform companies' software systems, in order to foster accountability not only with respect to algorithmic decision-making processes (which data rights legislation would not fix), but also to business operations more generally. Here we suggest a combination of company and platform or software audits conducted by teams of elected public officials and experts, as well as due-process protocols that grant gig workers the ability to appeal their deactivation and that ensure a speedy and fair arbitration or tribunal review process.

All these measures are ultimately geared toward increasing the operational costs of predatory platform businesses and thereby increasing the risks (which are also costs) of investing in these businesses. In this way, they are likely to have an upstream impact. Yet we could move further up the value chain and think of even more ambitious regulatory schemes that directly affect the operations of meta-platforms. One disincentivising measure is to increase capital gains tax on the sale of gig economy-related assets, which would make it costlier for a firm such as SoftBank to sell its shares of a platform company. A similar measure is to tighten financial regulations to raise the costs of investing in gig economy platforms, for instance, by more robustly taxing private equity transactions. As Rahman (2018, p. 249) has argued, 'such structural regulations would change the incentives in a way that makes the more problematic downstream practices less profitable and thus less likely'. Finally, following Rahman, we 'might impose antitrust-style limits on mergers and acquisitions', which could, for instance, block Uber's pending acquisition of Grubhub, or 'prevent the concentrated ownership over multiple ... platforms and related services into too few investor hands' (Rahman 2018, p. 249). This type of legislation would effectively make the data empire-building ventures of meta-platforms a great deal more onerous.

Policy Innovation and Public Investments

It is important to highlight here that private costs can be public gains. Building on expanded and intensified regulations targeting corporate (meta-)platforms, we should extend this radical ambition to the realm of public policy in order to foster novel and emerging platform-based initiatives, which can be funded by newly obtained tax revenues. Coopcycle, for instance, could hugely benefit from state subsidies, which would be more effective and sustainable than relying on crowdfunding or social impact investors whose support is usually contingent on particular deliverables. These subsidies could be managed by local governments, which would enable new public-private or common partnerships that encourage collaboration between platform cooperatives, their stakeholders or shareholders, and municipalities that would consequently be reinvigorated after years of budget cuts to social services. This would, however, require an openness to this type of collaboration on the part of the platform cooperativism movement, whose entrepreneurial and activist inclinations have so far resulted in a reticence to depend on (and report to) the state.

Furthermore, local governments could use new tax revenues to initiate 'public options' that 'would provide alternatives for users [including workers], while also exerting competitive pressure on otherwise dominant ... platforms, forcing those platforms to take seriously the need to provide services in a different way' (Rahman 2018, p. 249). One significant advantage of publicly initiated platforms, compared with the nationalisation of corporate platforms, is the ability to build new software architectures and platform ecosystems that are not rooted in exploitation or accumulation by dispossession. Public ownership of Deliveroo does not automatically entail a more equitable platform, given that the machine learning algorithms Deliveroo has developed are trained by data captured from a labour process that is specifically engineered for the purpose of revenue optimisation and worker subordination. The decisions that these algorithms execute are not in the public interest nor do they serve the common good, and corporate algorithmic systems should thus at the very least be de-programmed and re-programmed. It may be preferable, however, to build new publicly governed systems from the ground up.

Ultimately, new policies and platform-based initiatives should first and foremost be committed to the affirmation of public values on a regional, national and supranational level (Van Dijck et al. 2019). These values are vital for a platform society in which data exists as a public asset that can be leveraged by all stakeholders participating in its collective governance. Truly solving our puzzle demands that we imagine a future beyond the gig economy and platform capitalism, by embedding labour advocacy within a redistributive social policy framework and a broader transnational politics of social justice.

REFERENCES

Aloisi, A. (2015), 'Commoditized workers: case study research on labor law issues arising from a set of on-demand/gig economy platforms', *Comparative Labour Law and Policy Journal*, **37** (3), 653–90.
Alpeyev, P. (2019), 'What Uber's IPO tells us about SoftBank's big ride-hailing bet', *Bloomberg Online*, accessed 23 April 2021 at https://www.bloomberg.com/news/articles/2019-05-08/what-uber-s-ipo-tells-us-about-softbank-s-big-ride-hailing-bet.

Alpeyev, P., G. Tan, M. Davis and E. Huet (2019), 'SoftBank unveils $9.5 billion WeWork rescue, gets 80% stake', *Bloomberg Online*, accessed 23 April 2021 at https://www.bloomberg.com/news/articles/2019-10-23/wework-unveils-softbank -rescue-package-neumann-to-leave-board.

Browning, L. and E. Newcomer (2019), 'Uber created a $6.1B Dutch weapon to avoid paying taxes', *Accounting Today Online*, accessed 23 April 2021 at https://www .accountingtoday.com/articles/uber-created-a-61b-dutch-weapon-to-avoid-paying -taxes.

Butler, S. (2019), 'Deliveroo makes £232m loss despite doubling its operation', *Guardian Online*, 2 October, accessed 23 April 2021 at https://www.theguardian .com/business/2019/oct/02/deliveroo-makes-232m-loss-despite-doubling-its -operation.

Couldry, N. and U.A. Mejias (2019), 'Data colonialism: rethinking big data's relation to the contemporary subject', *Television and New Media*, **20** (4), 336–49.

Fraser, N. (2016), 'Expropriation and exploitation in racialized capitalism: a reply to Michael Dawson', *Critical Historical Studies*, **3** (1), 163–78.

Gallagher, C., W. Li and K. Gregory (2019), 'Restoring gig workers to power the collective potential of personal data portability', paper presented at the Privacy Law Scholars Conference, Amsterdam, October.

Gandini, A. (2019), 'Labour process theory and the gig economy', *Human Relations*, **72** (6), 1039–56.

Harvey, D. (2005), *The New Imperialism*, Oxford: Oxford University Press.

Hawkins, A. (2019), 'Uber lost over $5 billion in one quarter, but don't worry, it gets worse', *The Verge*, 8 August, accessed 23 April 2021 at https://www.theverge.com/ 2019/8/8/20793793/uber-5-billion-quarter-loss-profit-lyft-traffic-2019.

Holder, S. (2019), 'For ride-hailing drivers, data is power', *CityLab*, 22 August, accessed 23 April 2021 at https://www.citylab.com/transportation/2019/08/uber -drivers-lawsuit-personal-data-ride-hailing-gig-economy/594232/.

Langley, P. and A. Leyshon (2017), 'Platform capitalism: the intermediation and capi-talisation of digital economic circulation', *Finance and Society*, **3** (1), 11–31.

Lyft, Inc. (2019), 'Lyft form S-1 registration statement', US Securities and Exchange Commission (initial public offering filings), Washington, DC, accessed 23 April 2021 at https://www.sec.gov/Archives/edgar/data/1759509/000119312519059849/ d633517ds1.htm.

Madsbjerg, S. (2017), 'It's time to tax companies for using our personal data', *New York Times*, 14 November, accessed 23 April 2021 at https://www.nytimes.com/ 2017/11/14/business/dealbook/taxing-companies-for-using-our-personal-data.html.

Medieros, J. (2019), 'How SoftBank ate the world', *WIRED*, accessed 23 April 2021 at https://www.wired.co.uk/article/softbank-vision-fund.

Meijerink, J. and A. Keegan (2019), 'Conceptualizing human resource management in the gig economy', *Journal of Managerial Psychology*, **34** (4), 214–32.

Mezzadra, S. and B. Neilson (2017), 'On the multiple frontiers of extraction: excavat-ing contemporary capitalism', *Cultural Studies*, **31** (2–3), 185–204.

Nussey, S. (2020), 'SoftBank plans $41 billion of asset sales to expand buyback and cut debt', *Reuters*, accessed 23 April 2021 at https://uk.reuters.com/article/us-softbank -group-buyback/softbank-plans-41-billion-of-asset-sales-to-expand-buyback-and -cut-debt-idUKKBN21A0F2.

Ongweso, E. (2020), 'SoftBank's new strategy: screw over startups not investors', *Vice*, accessed 23 April 2021 at https://www.vice.com/en_ca/article/y3mdej/softbanksnew -strategy-screw-over-startups-not-investors.

Prassl, J. (2018), *Humans as a Service: The Promise and Perils of Work in the Gig Economy*, Oxford: Oxford University Press.

Rahman, K. (2018), 'Regulating informational infrastructure: Internet platforms as the new public utilities', *Georgetown Law Technology Review*, **2** (2), 234–51.

Rosenblat, A. and L. Stark (2016), 'Algorithmic labor and information asymmetries: a case study of Uber's drivers', *International Journal of Communication*, **10**, 3758–84.

Sadowski, J. (2019), 'When data is capital: datafication, accumulation, and extraction', *Big Data and Society*, **6** (1), 1–12.

Sadowski, J. (2020), 'The Internet of landlords: digital platforms and new mechanisms of rentier capitalism', *Antipode*, **52** (2), 562–80.

Shapiro, A. (2017), 'The urban stack. A topology for urban data infrastructures', *Tecnoscienza*, **8** (2), 61–80.

Sherman, L. (2019), 'WeWork's failure is SoftBank's day of reckoning', *WIRED*, accessed 23 April 2021 at https://www.wired.com/story/weworks-failure-is -softbanks-day-of-reckoning/.

Srnicek, N. (2017), *Platform Capitalism*, Cambridge: Polity Press.

Stewart, A. and J. Stanford (2017),' Regulating work in the gig economy: what are the options?', *Economic and Labour Relations Review*, **28** (3), 420–37.

Tomassetti, J. (2016), 'It's none of our business: the postindustrial corporation and the guy with a car as entrepreneur', paper presented at the meeting of the Society for the Advancement of Socio-Economics, Berkeley, CA, June.

Uber Technologies, Inc. (2019), 'Registration statement under the Securities Act of 1933', US Securities and Exchange Commission (initial public offering filings), Washington, DC, accessed 23 April 2021 at https://www.sec.gov/Archives/edgar/ data/1543151/000119312519103850/d647752ds1.htm.

Van Dijck, J,. D. Nieborg and T. Poell (2019), 'Reframing platform power', *Internet Policy Review*, **8** (2), 1–18.

Van Doorn, N. (2017), 'Platform cooperativism and the problem of the outside', *Culture Digitally*, accessed 23 April 2021 at http://culturedigitally.org/2017/02/ platform-cooperativism-and-the-problem-of-the-outside/.

Van Doorn, N. and J.Y. Chen (forthcoming), 'Odds stacked against workers: labor process gamification on Chinese and American food delivery platforms', accessed 23 April 2021 at https://admin.platformlabor.net/output/labor-process-gamification -china-us-food-delivery-platforms/Odds%20Stacked%20Against%20Workers_pre -pub.pdf.

Van Doorn, N., F. Ferrari and M. Graham (2020), 'Migration and migrant labor in the gig economy: an intervention', *SSRN Electronic Journal*, unpublished paper, accessed 23 April 2021 at https://papers.ssrn.com/sol3/papers.cfm?abstract_id= 3622589.

Veen, A., T. Barratt and C. Goods (2020), 'Platform-capital's "app-etite" for control: a labour process analysis of food-delivery work in Australia', *Work, Employment and Society*, **34** (3), 388–406.

7. Online labour platforms, human resource management and platform ecosystem tensions: an institutional perspective

Anne Keegan and Jeroen Meijerink

THE SOCIAL PROBLEM: HUMAN RESOURCE MANAGEMENT TO CONTROL FREELANCE GIG WORKERS

This chapter addresses the tensions arising from institutional complexity that are associated with the use of human resource management (HRM) activities by many online labour platforms (OLPs). Online labour platforms use HRM activities to control freelance gig workers and those who request their services. The use of HRM activities by OLPs generates institutional complexity because OLPs circumvent the employment relationship which is central to HRM as a disciplinary field. Extant HRM thinking is founded on the idea that HRM activities, such as staffing, training, appraisal, compensation and job design, serve to shape and manage the employment relationship (Lepak and Snell 1999; Nishii and Wright 2008; Tsui et al. 1997). Human resource management scholars look on, bemused, at current developments linked with work in OLPs in the gig economy. As intermediaries between supply and demand for labour, most OLPs view themselves as not employing gig workers, who instead are regarded as independent contractors providing services in the online market(s) that OLPs create (Koutsimpogiorgos et al. in press; Veen et al. 2020; Wood et al. 2019). Although gig work mainly takes place outside the confines of an employment relationship, most gig workers are nevertheless subject to a range of HRM activities, including: recruitment, selection, appraisal, compensation and job design (Duggan et al. 2020; Meijerink and Keegan 2019). This new phenomenon – HRM activities without employment relationships – is the platform puzzle we examine in this chapter.

According to Meijerink and Keegan (2019), a platform ecosystem perspective best explains the use of HRM activities in the non-employment context that characterises most OLPs. For the purposes of this chapter, we define platform ecosystems as the aggregate of gig workers, requesters and online labour platform firms who are semi-autonomous, yet interdependent actors (Breidbach and Brodie 2017; Jacobides et al. 2018). Gig workers and requesters (organisations and/or consumers that wish to outsource a fixed-term task) are interdependent. A limited supply of gig workers poses challenges for requesters to outsource activities, while low levels of labour demand from requesters mean gig workers' opportunities for generating income are compromised. Moreover, OLP firms' revenues are based on charging fees for intermediation services; therefore, they are also dependent on gig workers' and requesters' willingness to (continue) transact(ing) with each other in online marketplaces created by OLPs (Jacobides et al. 2018). Human resource management activities are instrumental in retaining and, even, locking gig workers and requesters into the platform ecosystem (Meijerink and Keegan 2019).

All this implies that, despite the absence of employment relationships, gig workers are subject to control by HRM activities initiated by OLPs. These activities are traditionally central to the standard employment relationship (Lepak and Snell 1999; Tsui et al. 1997), which creates a puzzle, as most platforms deny the existence of an employment relationship and refute claims that there should be one (Rosenblat 2018). For example, in their study of OLPs, Duggan et al. (2020, p. 116) hold that 'the means by which gig organisations view their workers has been controversial'. Likewise, Kuhn and Maleki (2017, p. 183) suggest that '[t]he rapidly growing number of people who find work via online labour platforms are not employees, nor do they necessarily fit traditional conceptualisations of independent contractors, freelancers, or the self-employed. The ambiguous nature of their employment status and its implications for worker well-being have attracted substantial controversy'. The key platform economy puzzle that this creates is the confusion about whether gig workers are employed by platform firms, or whether they are freelancers as platform firms claim. It is precisely this puzzle, sparked by OLPs' use of HRM activities to control gig workers, that we focus on in this chapter.

We explain that many OLPs adopt HRM activities to control freelancers and coordinate platform ecosystems, and this creates tensions that can be conceptualised as institutional complexity, that is, complexity arising from alignment by organisations to contradictory institutional logics (Greenwood et al. 2011). In line with the work of Frenken et al. (2020) and Meijerink et al. (in press), the logics in question, we argue, are those of the corporation and the market. We argue that these tensions are observed by societal stakeholders or institutional players residing outside, but relevant to, the platform ecosystem, and are used to challenge the legitimacy of OLPs. Also, OLPs may take the opportunity afforded by the currently fragmented nature (Vermeulen et al. 2014) of the emerging gig economy to establish novel HRM activities that are less likely to occur within standard employment relationships. Accordingly, we examine how OLPs seek to address institutional complexity by controlling gig workers, while simultaneously disavowing they are employers to gig workers, which ultimately (re)produces the platform puzzle of HRM for freelance gig workers. We conclude with directions for future research on the institutional complexity associated with the use of HRM activities for controlling gig workers.

THE SCIENTIFIC PROBLEM: THE NEED FOR AN INSTITUTIONAL COMPLEXITY LENS

The issues around employment status have become more complicated in recent years and, with the rise of platform-enabled gig work, this is likely to intensify. Hepple's writing at the end of the 1980s already held that

> [w]hat has changed over the past few decades has been the pattern of work, from full-time employment in the 'core' of the labour market towards part-time and temporary work (particularly among women) and a variety of forms of self-employment or of work in the twilight area between employment and self-employment. (Hepple 1986, p. 73)

Arguably, that twilight area is becoming far greater with the advent of OLPs, and the controversy over employment status of gig workers (Aloisi 2016; De Stefano 2015), which has been developing for decades is likely to grow. In our view, this controversy is the result of the HRM-enabled control that OLPs exercise over gig workers who, as independent contractors, are supposed to be autonomous in their work activities (Meijerink et al. in press).

We are not the first to write about how OLPs create tensions between control and autonomy. For instance, multiple authors refer this to as the autonomy paradox between workers' desire for autonomy that is negated by the OLPs' need for control (Kuhn and Maleki 2017; Möhlmann and Zalmanson 2017; Veen et al. 2020; Wood et al. 2019). Moreover, OLPs are criticised for promising autonomy, freedom and flexibility to gig workers, while restraining and controlling gig worker behaviour by means of online appraisal schemes, algorithmic dispatching or predetermined compensation schemes that harm gig workers' interests (Kuhn and Maleki 2017). For example, in research on the remote provision of digital services mediated by OLPs, Wood et al. (2019, p. 56) argue that '[a]lgorithmic management techniques tend to offer workers high levels of flexibility, autonomy, task variety and complexity. However, these mechanisms of control can also result in low pay, social isolation, working unsocial and irregular hours, overwork, sleep deprivation and exhaustion'.

To conceptualise these tensions, a majority of gig economy researchers (implicitly) draw on labour process theory (Edwards 1975; Smith 2015) to examine how OLPs extract and control gig workers' labour effort and how gig workers resist to (re)gain autonomy (Kellogg et al. in press; Veen et al. 2020; Wood et al. 2019). Although this has helped to shed light on gig workers' coping mechanisms to ensure freedom and flexibility – such as circumvention and manipulation of control (Jarrahi and Sutherland 2019) – little is known about how OLPs themselves balance the tension between control and freedom or flexibility. This is important since some of the OLPs' HRM practices – For example, staffing, compensation, training, appraisal and job design – serve to control gig workers, while others – such as autonomy on where, how, when and for whom to work – offer gig workers the freedom and flexibility that is supposed to go together with their freelance status (Meijerink et al. in press). It is precisely the combined use of control- and freedom-enhancing HRM activities that sparked debates on the employment status of gig workers and the legitimacy of OLPs. Moreover, OLPs also control the actions of requesters (for instance, restaurants that work with Deliveroo riders) by means of HRM activities such as online appraisal schemes, selection of requesters, admitting requesters access to online marketplaces, and sanctions such as the algorithmic restriction of the number of consumers that requesters can serve (Meijerink and Keegan 2019; Rosenblat 2018). These activities go against the independent status of requesters, as independent businesses, which is not necessarily captured by labour process theory. To better understand how OLPs balance control and autonomy by means of HRM activities, researchers such as Frenken et al. (2020) and Meijerink et al. (in press) propose institutional theory as a useful theoretical lens. Specifically, the concept of institutional complexity can be fruitful.

Following Vermeulen et al. (2014), institutional complexity arises as organisations face diverse influences arising from the coming together of different institutional logics owing to unprecedented levels of environmental turbulence, and fading boundaries between firms and industries (Greenwood et al. 2011). Institutional logics are 'the socially constructed, historical patterns of material practices, assumptions, values, beliefs, and rules by which individuals produce and reproduce their material subsistence, organise time and space, and provide meaning to their social reality' (Thornton and Ocasio 1999, p. 804). They represent ideal-type representations of an organisation's institutional field which shape the actions of organisations by explicating which goals and activities are seen as legitimate (Thornton 2004; Thornton and Ocasio 1999). Careful responses are required to secure legitimacy from different sources while also ensuring performance and survival of the organisation, all of which requires negotiation and trade-offs between institutional logics (Greenwood et al. 2011). Institutional logics, such as the market and corporation logics, are supported by macro-institutions, such as social dialogue and worker rights (in relation to corporation) and antitrust law (in relation to market logic). Online labour platforms create institutional complexity when they attempt to combine different logics such as the corporation logic (that is, control and coordination) with the market logic (autonomy, freedom and flexibility) (Frenken et al. 2020). It is the use of HRM activities by OLPs, for coordinating and controlling the efforts of both gig workers and requesters for gig labour, which creates tension between the logics of the market and corporation (Meijerink et al. in press).

HRM ACTIVITIES AS SOURCES OF INSTITUTIONAL COMPLEXITY IN ECOSYSTEMS

HRM Activities, Freelancing and the Market Logic

Online labour platforms, gig workers and requesters are rooted, to some extent, in the market logic. Under the market logic, profit-making is seen as a legitimate goal which can be achieved through free, unregulated competition (Thornton 2004). By aligning with the market logic, OLPs claim to be neutral intermediaries, offering software as a service, and simply facilitating online marketplaces where gig workers, as one-person businesses, can engage in free and unregulated competition for making profits (Rosenblat 2018; Schmidt 2017). For some platforms, for example, Helpling, Fiverr and Upwork, this market logic alignment is highly consistent with the way the platform treats its workers, that is, as freelancers. Instituting a market logic aligns well with the OLPs' business model and, more specifically, making profits. By working with freelance workers, OLPs try to avoid the (labour) costs associated with the standard employment relationship, such as pensions, social security, paid holidays and insurances (Aloisi 2016; Daskalova, 2018).

Online labour platforms may use several HRM activities that enable them to align with market logics. As noted by Meijerink et al. (in press) these may include staffing and autonomy. By setting few selection criteria and easily granting access to the online platform, new gig workers face low barriers to join the online marketplaces that OLP's create. For instance, applying to work for Deliveroo is decidedly uncomplicated and easy as selection is very liberal. All riders who meet basic criteria (for instance, have a suitable smartphone and can legally work) are accepted (Meijerink et al. in press; Veen et al. 2020). Moreover, by granting autonomy and freedom to gig workers to decide when, where, how, for whom and against which price to work, OLPs further align with market logics. Offering autonomy enables gig workers, at least in theory, to create (more) profit as they have the freedom, as entrepreneurs, to make choices themselves, such as working for those clients willing to pay a premium price for a gig worker's service, by competing on the basis of price and/or by competing on the basis of service (quality).

Although platforms disavow that they employ workers, and claim simply to facilitate a digital marketplace where labour providers and labour requesters can interact (Rosenblat 2018), this market logic is contradicted by the hands-on involvement of some platforms in the interactions between gig workers and requesters (Frenken et al. 2020). Specifically, the autonomy that gig workers should enjoy according to the market logic, is offset by HRM activities such as platform-set compensation schemes, algorithmic matchmaking between gig workers and requesters, and performance appraisal (Meijerink and Keegan 2019). These HRM practices are needed to coordinate platform ecosystems which reinforce the corporation logic of control, but which contradict the market logic of free and unregulated competition among gig workers that OLPs also propagate.

HRM Activities, Platform Ecosystems and the Corporation Logic

Online labour platforms, gig workers and requesters are also rooted, to some extent, in the corporation logic. Under the corporation logic, market share and revenue growth are seen as legitimate goals that are achieved by means of control and coordination (Thornton 2004). Despite working with stakeholders, that is, gig workers and requesters, who are supposed to be independent (that is, market logic), there is a clear need for OLPs to exercise coordination and control over gig workers and requesters. Some platforms align tightly with the corporation logic, and seek high levels of control and coordination over their workers, including examples such as Foodora, who employ their workers. The need for coordination and control, however, is shared to some degree by all platforms, and follows from that gig workers, requesters and OLPs making up a platform ecosystem of interdependent, yet semi-autonomous, actors (Meijerink and Keegan 2019). A platform ecosystem refers to a group of interacting, yet semi-autonomous entities – including at least gig workers, requesters and the OLP – that depend on each other's activities and therefore are hierarchically controlled (Jacobides et al. 2018; Wareham et al. 2014).

Interdependencies in platform ecosystems
Managing the interdependencies among the three ecosystem actors is important for ensuring smooth transactions in online marketplaces and generating revenues. We explain this using the example of the Deliveroo platform ecosystem. First, gig workers and requesters are mutually dependent on each other regardless of whether OLPs align more to market than to corporation logics for managing workers, or vice versa. In the example of the Deliveroo ecosystem, when many meal-deliverers leave the ecosystem (that is, stop using the online

platform to acquire meal delivery orders), there is limited value created for restaurants and consumers as meals are not delivered or are delivered too late. The same is true, however, for a platform such as Foodora, who employ their riders.

Moreover, meal deliverers have to ensure that meals are delivered appropriately (that is, keep a good temperature, items are kept separate, and so on) to avoid damaging the reputation of the restaurant and to ensure consumers receive high-quality meals. Gig workers (and restaurants) are also dependent on consumers: when consumers stop ordering food via the Deliveroo platform, both restaurants and gig workers will not reap benefits as insufficient earnings can be made. Moreover, gig workers are dependent on restaurants: when restaurants take too long preparing an order, it means gig workers end up wasting time that could have been spent on delivering extra orders and thus, generating more earnings (Meijerink et al. in press; Veen et al. 2020).

The value to OLPs also depends on the ongoing contribution of gig workers and requesters: when many gig workers, restaurants and/or consumers decide to leave the Deliveroo ecosystem, it means there are limited ongoing transactions in Deliveroo's online marketplace, lowering Deliveroo's capacity to capture fees from transactions. Actors within platform ecosystems should therefore be considered complementary and mutually dependent: the actions of one ecosystem actor have implications for other ecosystem actors, while also impacting on value other actors can derive from participating in the ecosystem (Adner 2017; Breidbach and Brodie 2017; Wareham et al. 2014).

It is in the interest of the OLP to control gig workers and requesters, as well as to coordinate the interdependencies among all ecosystem actors, as this ultimately increases the revenues accruing to the OLP (that is, corporation logic). Specifically, it is in the platform firms' interests to ensure liquidity in transactions, so that gig workers and requesters continue to transact, and so the OLP can charge a monetary fee based on these transactions. To ensure ongoing transactions, both gig workers and requesters should derive benefits from transacting with one another and the OLP, as this ensures their willingness to transact via the online platform.

Although we have illustrated these points with reference to Deliveroo, and the ecosystem it has created, which adopts a market aligned institutional logic to manage its relationship with freelance workers, the need to ensure ongoing and smooth transactions between all actors in the ecosystem holds also for platform ecosystems with strong corporation-logic alignments for workers, such as Foodora, who employ their riders. This is because the benefits of platform ecosystems depend, at least partly, on the ability of the OLP to create network effects (Katz and Shapiro 1994), as explained next.

Network effects in platform ecosystems

Network effects occur when the 'the value of membership to one user is positively affected when another user joins and enlarges the network' (Katz and Shapiro 1994, p. 94). Network effects occur in OLPs where platforms grow the number of workers in the online marketplace in order to attract additional requesters and generate more demand for their intermediation services (from individuals and businesses) which leads to more workers as they are attracted by the work opportunities which enlarge with the addition of new requesters (for instance, individuals ordering food or hailing rides, or restaurants attached to the platform supplying work to meal deliverers). These network effects can bring great power and status to platforms that have gained such a large market share (as a legitimate goal under the corporation logic) that they can become almost too big to fail.

Platforms try to create network effects by ensuring that, as requester numbers grow, sufficient numbers of workers sign up to service the growing number of clients. Platforms may therefore offer generous rates in early days as they grow the platform and scale up operations, which is legitimate under the corporation logic of growth (Frenken et al. 2020). However, balancing network effects, and in effect the interdependencies among platform ecosystem actors, is difficult. Oversupply of workers compared to demand from requesters will probably damage worker motivation to join the platform owingto worries about insufficient work and competition for work. If this happens, the value of platform participation by workers drops, and this can lead, in turn, to less value for clients (individual clients but also business, such as restaurants) from their participation in the platform. Too many requesters compared with workers may lead to delays in deliveries (for food delivery platforms such as Doordash in New York), rides (for ride hailing firms such as Ola in India or Didi in China), shelf stocking (for incidental shiftwork platforms such asWonolo), or pet care (for dog walking platforms such as WAG!), leading to lower motivation on the part of requesters which in turn leads to less value of participation for workers. As shown by Meijerink et al. (in press), the balancing of supply or demand that OLPs try to achieve is difficult, particularly before a platform has achieved a core or monopoly position, and while it is still competing with other platforms for the same workers or requesters who are also navigating the tensions of balancing supply and demand. In this respect, and particularly in the early phases of their existence, OLPs are strongly rooted in the corporation logic under which growth is seen as a legitimate goal achieved by means of control and coordination. These forms of control and coordination are needed to manage the interdependenices among platform ecosystem actors and to create network effects. Therefore, both gig workers and requesters are subject to HRM activities that afford control and coordination (Meijerink and Keegan 2019).

HRM practices for workers that institute the corporation logic
Online labour platforms use HRM practices to control gig workers in such
a way that gig workers co-create value with requesters and network effects
are created (Cassady et al. 2018; Ellmer and Reichel 2018; Meijerink and
Keegan 2019). For instance, they need to attract workers to their platform
(recruitment and selection), and they need to design their jobs (job design)
so that the worker behaves (discipline and incentives) in a way that enables
platform growth which is linked with client and requester satisfaction to
continue to transact (Kuhn and Maleki 2017). They need to manage worker
performance (performance management, discipline and appraisal) in line with
platform growth aspirations in order to meet requesters' needs. They need
to reward and discipline workers to ensure platforms can continue to grow
by giving clients the service quality they expect in line with what platforms
promise them (Rosenblat 2018; Veen et al. 2020). The promises made differ
from platform to platform, but include promises that, for instance, taxi drivers
will be trustworthy and prompt, that cleaners will be competent and respectful
of client privacy, and that there are enough workers to meet client demand
for prompt service so that rides, dog walks and shelf-stocking occur promptly
and correctly. Appraisal schemes may also ensure a sufficient supply of gig
workers. The main form of performance appraisal in OLPs is the system
of online performance ratings. Lack of portability of ratings can provide
a disincentive for workers to move platforms, particularly when working on
platforms such as Mturk, Deliveroo and Upwork where ratings are coupled
with opportunities to acquire work and, even, access better paid gigs (Jarrahi
and Sutherland 2019; Kellogg et al. in press). Having a high rating serves as
a signal when seeking work and staying with the platform and working con-
tinuously or regularly are heavily incentivised by the performance appraisal
processes used. To attract clients or requesters to platforms, it is common for
OLPs to design payment schemes that include gig workers working on speculation, and producing work for clients which remains unpaid until the client
approves the work, which they sometimes fail to do (Jarrahi and Sutherland
2019). Moreover, OLPs may determine the compensation of gig workers to
ensure requesters' are willing to transact with a gig worker (Kuhn and Maleki
2017). As gig workers get locked-in by means of online rating schemes, OLPs
may decide to further lower gig worker income to attract additional requesters
to the platform (Rosenblat 2018).

HRM practices for requesters that institute the corporation logic
Given the interdependent nature of activities in platform ecosystems, requesters are also subject to HRM activities. In line with the corporation logic of
control, OLPs use HRM activities to steer requesters' behaviours (Meijerink
and Keegan 2019). In cases such as Foodora, the use of HRM activities will

be similar to those used in non-platform companies which, similar to Foodora, have an employment relationship with workers. However, even for platforms that do not formally employ workers, using standard employment contracts, we also find HRM activities for requesters. For example, Uber requires taxi drivers to rate the performance of requesters in order to create trust and build perceptions of personal safety on the part of drivers during ride-hailing services (Rosenblat et al. 2017). These ratings can be used by Uber to deactivate ride-hailers for inappropriate behaviour that is reported by Uber drivers through the online application (app). Online labour platforms may use complex algorithms that consider requesters' ratings by the worker. In Deliveroo, the algorithm can be programmed with feedback on whether restaurants have orders ready on time, to ensure that riders are not wasting time waiting on orders to be prepared (Meijerink et al. in press). In the event that restaurants do not comply with Deliveroo's standards, an algorithm implements sanctions that restrict the likelihood that consumers will order at the selected restaurant. This enables the conditions for ongoing engagement on both the labour supply and labour demand side (Kuhn and Maleki 2017). In line with this, requesters who frequently outsource a high number of activities via the online labour platform may be rewarded for this with discounts and reduced commission fees. Also, requesters may be subject to stringent selection activities. For instance, to ensure a balanced representation of cuisines, Deliveroo may decide not to admit (for example) additional fast-food restaurants when these are likely to dominate the platform (Meijerink et al. in press). Online labour platforms may recruit additional requesters by offering temporary price reductions to entice them in order to ensure that as platforms are growing, and gig workers are not put off by pressure on their 'wages' coming from requesters looking for better deals. These subsidies are designed to ensure both sides perceive they get enough value from transacting in the early stages of platform growth, even if such incentives prove unsustainable over time. By the time platforms have dominated and driven competitors out, becoming too big to fail, subsidies to requesters may disappear and payment terms for gig workers are prone to often multiple, sequential disadvantageous changes (Meijerink and Keegan 2019).

CONSEQUENCES OF INSTITUTIONAL COMPLEXITY FOR ECOSYSTEM ACTORS

The use of HRM activities by most OLPs for managing gig workers and requesters creates tensions between the market logic and corporation logic since most OLPs do not employ workers using standard employment relationships (Frenken et al. 2020; Meijerink et al. in press). Specifically, the implementation of autonomy practices reinforces the market logic in that they offer, at least in theory, the possibility for gig workers to decide where, when, how, for how and against which price they want to work for maximising profit. This is offset however by HRM activities such as performance appraisal, selection, pre-set compensation and algorithmic control that align with the corporation logic. Alignment with this logic implies growing the platform ecosystem to increase the OLP's market share and revenue by steering gig-worker (and requester) behaviour, which ultimately limits the freedom that freelancers should have under the market logic.

The tensions between the market and corporation logics manifest as institutional complexity. Institutional complexity is defined as the incompatibility of prescriptions, in what are seen to be legitimate goals and actions, that come from different institutional logics (Greenwood et al. 2011; Vermeulen et al. 2014). The contradictions between the market and corporation logic create incompatibilities that have consequences for OLPs, gig workers and requesters.

Consequences of Institutional Complexity for Online Labour Platforms

Owing to the institutional complexity associated with their use of HRM activities, OLPs run the risk of losing their legitimacy in the eyes of other (societal) stakeholders. Online labour platforms and their platform ecosystems are embedded within broader institutional contexts, or institutional ecosystems. These broader contexts comprise other actors, including unions, labour regulators, politicians and journalists. One or more platform ecosystems may encounter similar broader institutional contexts. For example, Uber and Lyft in the USA probably have similar broader contexts to confront in relation to journalists covering the gig economy from a taxi or ride-hailing perspective. It is the tension between the use of control-enhancing HRM activities (that is, corporation logic) for gig workers that are supposed to be independent (that is, market logic) which made journalists write critical reports about OLPs such as Uber and Deliveroo (Lieman 2018). Ultimately, this may result in OLPs losing their legitimacy among requesters, which may stifle the growth of OLPs' online marketplaces. We write 'may' for two reasons. The first is that, following Vermeulen et al. (2014), fragmented fields such as OLPs may allow actors (for example, platform firms) more freedom to decide how to align with different logics or use different logics flexibly, much like a toolkit. This may be a temporary issue, with such flexibility fading as the institutional context becomes less fragmented. Second, in relation to OLPs, the majority of consumers may not experience institutional complexity associated with HRM for freelancers or, if they do, may not perceive it as problematic. Consumers are likely to be strongly rooted in a market logic, as research has shown that consumers make use of OLPs and gig workers on the basis of price (Möhlmann 2015). Consumers are likely to be sheltered from the corporation logic of control that gig workers are subject to, since the use of HRM activities is opaque to consumers. Consumers will be unlikely to challenge the legitimacy of OLPs as long as the prices for gig-worker services and intermediation services that are charged to them remain below a particular threshold. This situation may change in the future. One reason is that a single OLP may come to dominate a selected market and increase intermediation service fees to requesters (and gig workers). Another reason is that consumers may become aware of tensions linked with OLP approaches to managing workers if tensions between autonomy and control of workers become salient at a broader societal level, or if the risks that the marketplace approach to work poses for gig workers is exposed to consumers. The onset of the COVID-19 crisis has arguably drawn greater attention to the precarious work conditions of many essential though low-paid workers and exposed work practices that are institutionally complex to greater scrutiny by the media, and the public.

In the meantime, other societal stakeholders use the institutional complexity of HRM for freelancers to challenge the legitimacy of OLPs. These include, but are not exclusive to, labour unions and politicians who have a basis to challenge OLPs on their tension-filled employment models where they claim not to employ people, but manage workers in ways consistent with employment. This has brought about an increasing number of high-profile court cases around the world that are based on the question of whether gig workers are falsely self-employed and should be reclassified as employees (Aloisi 2016; Dubal 2018; Frenken et al. 2020; Meijerink and Keegan 2019). In countries such as the Netherlands, Spain, the UK and the USA, court rulings have explicated that the use of HRM activities for freelance gig workers implies the presence of an employment contract, meaning that OLPs are employers to their workers. As an example, Deliveroo offers favourable access to shifts to those who work at peak times, thus undermining the freedom of riders to choose when to work by incentivising particular platform-favoured shifts (Meijerink et al. in press; Veen et al. 2020). It was because of this that a Dutch judge ruled that Deliveroo riders should be reclassified as employees. This and similar verdicts may have serious consequences for OLPs since they potentially lead to increased labour costs, making OLPs' business models less viable. Particularly in online markets where profit margins are small, increases in labour costs may cause OLPs to go out of business.

Consequences of Institutional Complexity for Workers

When HRM practices become a focal point for some of the growing tensions surrounding OLPs, these tensions may spill over to, and have consequences for, actors within platform ecosystems, most notably gig workers and requesters.

The alignment of platforms with the corporation logic can be seen in their orientation towards growth, scale and the creation of network effects. Owing to network effects, platform growth is self-reinforcing as the value of a platform increases with the number of workers and requesters (Frenken et al. 2020). Workers on platforms could also be seen as operating in line with a corporation logic. After all, they are (according to platform firms) self-employed, independent and the smallest of small businesses. However, these independent workers cannot collaborate with each other to set prices or counter the power of the platform since this is not permitted under competition law (that is, market logic) (Daskalova 2018). As corporations, platform workers also cannot grow and expand their business owing to the constraints of having only their own hours of labour to sell, with little possibility to expand that or to hire others (Frenken et al. 2020). The ratings systems of OLPs allow gig workers to build a reputation online. However, these online reputations are tied to the individual and not portable to other platforms or workers. Therefore, as platforms grow, they may effectively trap workers despite their freelance status. In effect, the corporation and market logic are misaligned for gig workers (Frenken et al. 2020), with the consequence being that gig workers cannot compete freely and only increase profit to the extent that they are capable of working longer hours that come at the expense of their physical, social and emotional well-being.

Another key result of the institutional complexity associated with platform-enabled gig work, is that power is not distributed equally in the platform due to the access each party has to information about the interactions. In line with the corporation logic of coordination, platforms see all transactions and can try to manipulate the other parties, in real-time, to control how clients and workers interact with each other (Kellogg et al. in press; Rosenblat 2018; Schmidt 2017; Veen et al. 2020). Workers and requesters (for instance, individuals or businesses) have little access to information on all interactions, pricing decisions and other information they might use to manage their own interests more effectively. This limits them in their entrepreneurial activities, which ultimately goes against the market logic of free competition. Again, the only way workers can increase earnings is by working longer hours rather than competing on the basis of price or service offerings.

The severity of these consequences will differ based on the dependence of workers on platforms, a condition that varies a great deal (Kuhn and Maleki 2017). Workers who are heavily dependent on platforms, and earn most of the income through platform work, react more to perceived injustices when platforms change conditions and disadvantage workers through new HRM practices that further reinforce tensions among institutional logics. Rate changes, or changes to how income is determined, are responded to more vigorously when gig work via platforms is more significant to someone's living standards (Kuhn and Maleki 2017; Rosenblat 2018). This variation explains the tendency of workers to respond differently to opportunities and constraints presented by the institutional context of OLPs and the tensions in the logics that these platforms create by having HRM without employment relationships.

SOLUTIONS TO INSTITUTIONAL COMPLEXITY OF HRM FOR GIG WORKERS

There are several solutions to the institutional complexity associated with HRM for controlling freelance gig workers. These reside at the level of the OLP or platform ecosystem as well as at the wider institutional level in which platform ecosystems are embedded.

Response Strategies by Online Labour Platforms and Within Platform Ecosystems

The most straightforward solution to institutional complexity involves OLPs opting to align with one institutional logic, and ceding the other. Oliver (1991) refers, for example, to the response strategy of acquiescence, including examples such as rule-obeying and norm acceptance. This would mean either retaining HRM activities and offering an employment contract to workers to align with the corporation logic (as in the case of Foodora or following the thrust of California Assemby Bill 5, 2019), or doing away with the majority of HRM activities, reinforcing the freelance status of gig workers, and aligning fully with the market logic including allowing freelance workers all of the autonomy a pure alignment with market logics implies. As simple as this solution is, it is unlikely that OLPs will adopt it as, for most platform firms, this undermines their business model (Meijerink et al. in press). Online labour platforms will probably resist an either/or response to institutional complexity as the consequence of this would be increased labour costs (ceding the market logic) or the inability to coordinate platform ecosystems to create network effects (ceding the corporation logic). We see the opposite occurring in many instances where OLPs, despite controlling gig workers and aligning with corporation logic, also simultaneosuly propagate the market logic by holding that gig workers value the freedom and flexibility that is associated with their independent status.

Instead of than favouring one logic or another, OLPs use a variety of other response strategies to address the institutional complexity they create. Oliver (1991, p. 152) refers, for example, to a strategy such as avoidance involving examples such as concealing. We observe that OLPs follow these avoidance strategies by developing novel (concealed) forms of HRM outsourcing and types of covert HRM implementation to achieve control over workers' activities, but to try and avoid or obfuscate the existence of employment relationships (Meijerink et al. in press). Even as OLPs respond to rules governing employment and the designation of workers, and try to avoid actions which constitute employment, they also deploy algorithmic control, gamification, obfuscation and distancing to enact control over workers while trying to avoid the semblance of an employment relationship. These strategies are enabled by information technologies and the marketplaces that OLPs create in what is the still emerging and thus fragmented field of gig work. Vermeulen et al. (2014, p. 79) argue that 'in fragmented fields, actors have more choice about which pressures they select for conformity (Quirke 2013), and they may even be able to undermine dominant logics by drawing on alternative minority logics (Durand and Jourdan 2012)'. Frenken et al. (2020) observe both growing tensions between OLPs and other actors involved in the gig economy, such as unions, while recognising that OLPs are able to take advantage of the still nascent state of the gig economy and the lack of clarity currently existing about gig work, and the responsibilities of OLPs for workers. They see evidence of 'rising tensions between gig-economy platforms, tax agencies, regulators and labour unions. These tensions have emerged as platforms have been able to partly neutralise the role of unions and state regulations due to unclear legal jurisdictions regarding labour rights and platform responsibilities' (Kenney and Zysman, 2016). Online labour platforms are able to take advantage of the emergent nature of gig work as well as weak efforts by regulators to force OLPs to align fully with corporate logics and responsibilities linked with traditional employment and employers. This aligns with Oliver's identification of the variation in organisational responses to institutional complexity which is dependent 'on the institutional pressures toward conformity that are exerted on organisations' (Oliver 1991, p. 151).

Responses by OLPs to institutional complexity, in the form of novel HRM practices, appear to be emerging dynamically over time as platforms navigate the tensions of new ways of working involving no employment relationship while simultaneously involving forms of control, motivation and discipline imposed by platforms to govern transactions between ecosystem actors (Meijerink et al. in press). From the perspective of HRM scholarship, OLPs are a new phenomenon giving rise to familiar HRM practices which are mobilised by platforms in new ways. Online labour platforms responses might best be seen in light of the currently weak level of 'institutional pressures' toward conformity (Oliver 1991) exerted by labour regulations, trade unions and other sources of institutional pressure in what is currently a fragmented field. The HRM practices evident in platform ecosystems are often attempts to reframe employment relationships and issues of control and power over workers through obfuscation. Oliver (1991) identifies an avoidance-based strategic response to institutional complexity, noting efforts by organisations to conceal their non-conformity to institutional norms in ways that call to mind the actions of OLPS in concealing their direction of workers which would normally lead to a designation of employee status. As a consequence of this avoidance strategy, the institutional complexity associated with HRM activities for gig workers, and the consequences this has for workers, will probably prevail until stronger institutional pressures to conform to traditional employment logics put pressure on platforms to change their approach to workers. Since it is in the interest of OLPs to exercise control over independent gig workers, it is very unlikely that the majority of OLPs will proactively address the institutional complexity they create. Instead, changes in the institutional context, and growing pressure from institutions of labour regulation, seem a more likely way that tensions between the market and corporation logics will be alleviated by clearer alignment by OLPs to one or the other.

Institutional-level Solutions to Tensions between the Market and Corporation Logic

While OLPs may be reluctant to favour one of the two institutional logics, recent changes on the institutional level lay the groundwork for a more distinct divide between the market and corporation logics. In many jurisdictions where OLPs establish themselves, changes in the institutional context present opportunities for workers to seek clarification on their status as self-employed or employed in order to try to improve their living standards when platform-enabled gig work is a main part of their income. The legal context in California at the time of writing, with the introduction of Assembly Bill 5 (2019),[1] provides an institutional context for workers that promotes OLPs as employers, not as technology service providers, and which strengthens the corporation institutional logic within which OLP activities might be framed. The California Assembly Bill 5 (2019) provides institutional support for gig workers seeking to elucidate whether or not platforms are their employers and if they can acquire rights and protection under that legislation, including rights to minimum wages, rights to collectively bargain over wages and conditions, freedom from discrimination, and payment for vacation days and absence due to illness or injury. The more heavily dependent workers are on one online labour platform, the more likely it is they may use the enabling nature of the institutional context to seek reclassification as workers or employed. Workers who do not rely solely or mainly on their work with a platform for their livelihood are less likely to seek to use

---

[1]   In November 2020, California Proposition 22 was approved, which classifies app-based workers as independent contractors. As a consequence, OLPs such as Uber, Lyft, DoorDash and Instacart do not need to comply with Assembly Bill 5.

regulations available to them and more likely to resist institutional pressures to be deemed employees instead of self-employed.

The drawback of this approach is that (representatives of) gig workers have to initiate court cases, and incur the costs of that, in order to be reclassified as workers, while leaving OLPs with few incentives to balance institutional pressures. This follows from whether workers are freelancers or employees being a matter of law. It is possible, however, to imagine solutions that motivate OLPs to make clearer distinctions between the market and corporation logics. One example is that gig workers are, by default, employed by the OLP. It would then be up to the OLP to convince a judge that workers should be reclassified as independent contractors (Risak 2017).

While this solutions require choosing between the market or corporation logic, solutions that combine both competing logics may also be feasible. The introduction of a third legal status of worker, as already exists in the UK, is one possibility. In these cases, gig workers remain independent (that is, market logic) while simultaneously enjoying some of the benefits that employees enjoy, such as a minimum wage, paid holidays or paid parental leave, all of which are aligned with the corporation logic. Although this solution may benefit gig workers, it also has its drawbacks as those who are currently employed may be reluctantly drawn into the worker category and thus lose the rights attached to the employment contract.

A more viable hybrid option is that implemented by Hilfr in Denmark. Hilfr is an OLP where freelance gig workers can sell housekeeping services to private households. Hilfr signed a collective agreement which holds that freelancers are offered several benefits that employees are entitled to, such as a minimum wage of €19 per hour, contribution to pension savings, and paid holidays and sick leave. Those who work more than 100 hours per year via the Hilfr platform can opt out[2] of these agreements. Similarly, the city of New York has instituted a rule to the effect that Uber also combines the market and corporation logic. In principle, Uber drivers are seen as freelancers, under the condition that Uber ensures that its drivers earn a minimum hourly rate when active on the Uber platform. This will bring about the institution of the corporation logic since Uber has to implement a system that controls drivers' access to the platform to avoid an oversupply of workers that come at a cost for Uber. Unless Uber manages to attract more requesters, and balance supply and demand for labour, this will cause a situation where gig workers will not have

---

[2]   Hilfr promised to pay minimum fees to workers that remain independent. In August 2020, the Danish Competition and Consumer Authority regards these minimum fees as in breach of competition law and has ordered Hilfr to stop paying these fees. For a detailed discussion on changes needed in competition law we refer to Chapter 4 in this volume.

unlimited and free access to Uber's online marketplace, which goes against the market logic of free and unregulated competition.

Finally, a solution can be conceived where contradictions between the corporation and market logics have fewer consequences for the social security of gig workers. In many jurisdictions, social security, such as paid holidays, pension planning and paid sick leave, are almost solely tied to the employment contract. As a consequence, independent contracts are not entitled to these benefits or have to pay for these themselves, which often proves to be very costly. A solution therefore would be to require every worker, irrespective of the contract he or she has, to contribute to and receive entitlements to social security benefits. Although this does not alleviate the tensions between the corporation and market logic, it would help to reduce some of the precarities currently associated with the institutional complexity of HRM activities that control and restrain gig workers, while denying them employment rights.

DIRECTIONS FOR FUTURE RESEARCH

Since changes in labour laws require time, or may create new tensions between institutional logics, it remains opportune for researchers to study the institutional complexity associated with HRM for freelance workers. We see several avenues for future research that provide fertile ground for new HRM insights.

First, although the majority of, if not all, OLPs that work with independent contractors combine the competing logics of the corporation and the market, some have to ensure compliance with more than two logics. For instance, OLPs such as Care.com, DoctorOnDemand.com or GigNow work with gig workers whose activities and identities are strongly embedded (such as doctors, teachers, nurses and accountants) in the professional logic (which emphasises reputation as legitimate outcome through the accumulation of expertise). To ensure compliance with this logic, the provision of training is of utmost importance. Online labour platforms that face institutional pressures associated with the professional logic may find it challenging to offer training services to professional gig workers, as this may imply control that goes against the free-market logic that OLPs propagate. The same would count for OLPs such as SitterCity.com (that is, families hiring a babysitter to care for children in the family home) which is likely to be strongly rooted in the family logic that legitimises loyalty. Online labour platforms may institute performance criteria for appraising and rewarding gig workers that induce gig worker loyalty to families, but that again may go against the logic of the market. Given the likelihood that OLPs may have to comply with a variety of institutional logics, we call for future studies that examine how OLPs, through their use of HRM activities, combine more than two competing institutional logics, and how societal stakeholders respond to this, and the consequences for platform legitimacy.

Second, institutional complexity probably differs across OLPs. This may depend on the degree to which OLPs are able to rely on input control, such as the selection of gig workers as well as the degree to which gig workers (can) set their own prices. Online labour platforms such as Fiverr, Upwork and Toptal rely on algorithmic selection to decide which independent contractors are admitted access to the online marketplace. This form of input control, where gig workers are selected on the basis of their competences and work experience, increases trust and may lower the need for OLPs to rely on behavioural and bureaucratic controls associated with corporation logics. This would pertain particularly to OLPs that grant requesters the autonomy to select gig workers themselves, instead of being assigned a worker by the OLP, as this reinforces the logic of the market. By comparison, platforms such as Deliveroo and Uber are more likely to experience institutional complexity. Owing to the relatively standardised nature of the gig work performed, relying on stringent input controls related to selection makes little sense. This, however, would require more behavioural control of freelancers by means of algorithms for performance appraisal that reinforce the competing logic of the corporation. Moreover, there is little time for Deliveroo riders or Uber drivers to negotiate a price with individual requesters, owing to the on-demand and on-the-spot nature of meal deliveries and taxi rides. This may explain why OLPs determine compensation schemes for independent contractors, which clearly goes against the market logic of free, unregulated competition. As these examples show, the institutional complexity associated with HRM for freelance gig workers probably differs across OLPs. Therefore, we call for comparative studies that examine the use of HRM activities for different types of platform workers and how OLPs' and societal stakeholders' respond to the puzzles created by contradicting logics across different platforms.

Third, the saliency of institutional complexity probably differs across OLPs. Meijerink et al. (in press) show that in the Netherlands, Deliveroo attracted more criticism from societal stakeholders for combining contradictory logics than Uber Eats does. These difference were attributed to Uber Eats working with independent contracts ever since it started its Dutch operations, while Deliveroo changed from an employee model to a freelancer model for its meal deliverers (Meijerink et al. in press). This change made visible the shift from a corporate logic only (that is, riders being employed by Deliveroo) to a mixture of the market logic and the corporation logic (that is, use of control-enhancing HRM activities for freelance meal deliverers). Given these examples, we see a need for empirical studies that adopt a longitudinal perspective to examine dynamics in institutional complexity across the lifespan of OLPs.

Finally, an increasing number of incumbent organisations (and requesters) are setting up their own online labour markets. This includes temporary agencies that create online platforms where requesters can hire the temporary agency's employees. Unless these agencies continue to work with employees, there will be little institutional complexity at play while they remain fully rooted in the corporation logic (and, for example, only use the online platform as an e-commerce tool). There are examples of incumbent organisations that do establish online labour platforms where gig workers sell their services. For instance, Ernst and Young recently established the GigNow platform where freelancers can join projects that Ernst and Young is performing at one of their client firms. A similar type of OLP is the Talent Exchange platform run by PriceWaterhouseCoopers. Online labour platforms such as GigNow and Talent Exchange are likely to be sheltered from the general public as they are part of a bigger organisation that offers services to corporate clients instead of to private consumers. It is unclear which HRM activities these incumbent-run OLPs deploy and whether this creates tensions between contradictory institutional logics. Therefore, we call for research that explores the nature of these HRM activities and whether the corporations that these OLPs are embedded in create enabling affordances for addressing institutional complexity that stand-alone OLPs such as Uber, Deliveroo or Upwork do not enjoy.

REFERENCES

Adner, R. (2017), 'Ecosystem as structure: an actionable construct for strategy', *Journal of Management*, **43** (1), 39–58.

Aloisi, A. (2016), 'Commoditized workers: case study research on labor law issues arising from a set of on-demand/gig economy platforms', *Comparative Labor Law and Policy Journal*, **37** (3), 653–90.

Breidbach, C.F. and R.J. Brodie (2017), 'Engagement platforms in the sharing economy: conceptual foundations and research directions', *Journal of Service Theory and Practice*, **27** (4), 761–77.

Cassady, E.A., S.L. Fisher and S. Olsen (2018), 'Using eHRM to manage workers in the platform economy', in J.H. Dulebohn and D.L. Stone (eds), *The Brave New World of eHRM 2.0*, Charlotte, NC: Information Age, pp. 217–46.

Daskalova, V. (2018), 'Regulating the new self-employed in the Uber economy: what role for EU competition law', *German Law Journal*, **19** (3), 461–508.

De Stefano, V. (2015), 'The rise of the just-in-time workforce: on-demand work, crowdwork, and labor protection in the gig-economy', *Comparative Labor Law and Policy Journal*, **37** (1), 471–504.

Dubal, V.B. (2018), 'Employment law: the employee vs. independent contractor dichotomy', *The Judges' Book*, **2** (1), 51–8

Duggan, J., U. Sherman, R. Carbery and A. McDonnell (2020), 'Algorithmic management and app-work in the gig economy: a research agenda for employment relations and HRM', *Human Resource Management Journal*, **30** (1), 114–32.

Durand, R. and J. Jourdan (2012), 'Jules or Jim: alternative conformity to minority logics', *Academy of Management Journal*, **55** (6), 1295–315.

Edwards, R.C. (1975), 'The social relations of production in the firm and labor market structure', *Politics and Society*, **5** (1), 83–108.

Ellmer, M. and A. Reichel (2018), 'Crowdwork from an HRM perspective – integrating organizational performance and employee welfare', Working Paper 1, University of Salzburg.

Frenken, K., T. Vaskelainen, L. Fünfschilling and L. Piscicelli (2020), 'An institutional logics perspective on the gig economy', in I. Maurer, J. Mair and A. Oberg (eds), *Theorizing the Sharing Economy: Variety and Trajectories of New Forms of Organizing*, Bingley: Emerald.

Greenwood, R., M. Raynard, F. Kodeih, E.R. Micelotta and M. Lounsbury (2011), 'Institutional complexity and organizational responses', *Academy of Management Annals*, **5** (1), 317–71.

Hepple, B. (1986), 'Restructuring employment rights', *Industrial Law Journal*, **15** (1), 69–83.

Jacobides, M.G., C. Cennamo and A. Gawer (2018), 'Towards a theory of ecosystems', *Strategic Management Journal*, **39** (8), 2255–76.

Jarrahi, M.H. and W. Sutherland (2019), 'Algorithmic management and algorithmic competencies: understanding and appropriating algorithms in gig work', paper presented at the International Conference on Information Systems, Munich, 31 March–3 April.

Katz, M. L. and C. Shapiro (1994), 'Systems competition and network effects', *Journal of Economic Perspectives*, **8** (2), 93–115.

Kellogg, K., M. Valentine and A. Christin (in press), 'Algorithms at work: the new contested terrain of control', *Academy of Management Annals*.

Kenney, M. and J. Zysman (2016), 'The rise of the platform economy', *Issues in Science and Technology*, **32** (3), 61–9.

Koutsimpogiorgos, N., J. van Slageren, A.M. Herrmann and K. Frenken (in press), 'Conceptualizing the gig economy and its regulatory problems', *Policy and Internet*.

Kuhn, K.M. and A. Malek (2017), 'Micro-entrepreneurs, dependent contractors, and instaserfs: understanding online labor platform workforces', *Academy of Management Perspectives*, **31** (3), 183–200.

Lepak, D.P. and S.A. Snell (1999), 'The human resource architecture: toward a theory of human capital allocation and development', *Academy of Management Review*, **24** (1), 31–48.

Lieman, R. (2018), *Uber Voor Alles*, Amsterdam: Business Contact.

Meijerink, J.G. and A. Keegan, A. (2019), 'Conceptualizing human resource management in the gig economy: toward a platform ecosystem perspective', *Journal of Managerial Psychology*, **34** (4), 214–32.

Meijerink, J.G., A. Keegan and T. Bondarouk (in press), 'Having their cake and eating it too? Online labor platforms and human resource management as a case of institutional complexity', *International Journal of Human Resource Management*.

Möhlmann, M. (2015), 'Collaborative consumption: determinants of satisfaction and the likelihood of using a sharing economy option again', *Journal of Consumer Behaviour*, **14** (3), 193–207.

Möhlmann, M. and L. Zalmanson (2017), 'Hands on the wheel: navigating algorithmic management and Uber drivers', paper presented at the Thirty-eighth International Conference on Information Systems, Seoul.

Nishii, L.H. and P.M. Wright (2008), 'Variability within organizations: implications for strategic human resource management', in D.B. Smith (ed.), *The People Make the Place: Dynamic Linkages Between Individuals and Organizations*, Mahwah, NJ: Lawrence Erlbaum Associates, pp. 225–48.

Oliver, C. (1991), 'Strategic responses to institutional processes', *Academy of Management Review*, **16** (1), 145–79.

Quirke, L. (2013), 'Rogue resistance: sidestepping isomorphic pressures in a patchy institutional field', *Organization Studies*, **34** (11), 1675–99.

Risak, M. (2017), *Fair Working Conditions for Platform Workers: Possible Regulatory Approaches at the EU Level*, Berlin: Friedrich-Ebert-Stiftung, accessed 19 April 2021 at https://www.fes-connect.org/reading-picks/fair-working-conditions-for -platform-workers/.

Rosenblat, A. (2018), *Uberland: How Algorithms Are Rewriting the Rules of Work*, Oakland, CA: University of California Press.

Rosenblat, A., K.E. Levy, S. Barocas and T. Hwang (2017), 'Discriminating tastes: Uber's customer ratings as vehicles for workplace discrimination', *Policy and Internet*, **9** (3), 256–79.

Schmidt, F.A. (2017), *Digital Labour Markets in the Platform Economy*, Berlin: Friedrich-Ebert-Stiftung, accessed 19 April 2021 at http://www.bollettinoadapt.it/ wp-content/uploads/2020/10/13164.pdf.

Smith, C. (2015), 'Continuity and change in labor process analysis forty years after labor and monopoly capital', *Labor Studies Journal*, **40** (3), 222–42.

Thornton, P.H. (2004), *Markets From Culture: Institutional logics and Organizational Decisions in Higher Education Publishing*, Palo Alto, CA: Stanford University Press.

Thornton, P.H. and W. Ocasio (1999), 'Institutional logics and the historical contingency of power in organizations: executive succession in the higher education publishing industry, 1958–1990', *American Journal of Sociology*, **105** (3), 801–43.

Tsui, A.S., J.L. Pearce, L.W. Porter and A.M. Tripoli (1997), 'Alternative approaches to the employee-organization relationship: does investment in employees pay off?' *Academy of Management Journal*, **40** (5), 1089–121.

Veen, A., T. Barratt and C. Goods (2020), 'Platform-capital's "app-etite" for control: a labour process analysis of food-delivery work in Australia', *Work, Employment and Society*, **34** (3), 388–406.

Vermeulen, P., C. Zietsma, R. Greenwood and A. Langley (2014), 'Strategic responses to institutional complexity', *Strategic Organization*, **12** (1), 79–82.

Wareham, J., P.B. Fox and J.L. Cano Giner (2014), 'Technology ecosystem governance', *Organization Science*, **25** (4), 1195–215.

Wood, A.J., M. Graham, V. Lehdonvirta and I. Hjorth (2019), 'Good gig, bad gig: autonomy and algorithmic control in the global gig economy', *Work, Employment and Society*, **33** (1), 56–75.

8. Multi-party working relationships in gig work: towards a new perspective

James Duggan, Ultan Sherman, Ronan Carbery and Anthony McDonnell

INTRODUCTION

Overview of Chapter

In this chapter,[1] we discuss the unique nature of working relationships in the gig economy, explore how these relationships develop with the involvement of multiple parties, and assess the inherent disruption to the traditional employment relationship. We focus exclusively on platform-enabled gig work, primarily that which is executed via smartphone applications (apps), which we refer to as app-work. To avoid confusion, we outline the following terminology used throughout the chapter:

- We refer to the individual tasks completed by gig workers as 'gigs'.
- We refer to the independent workers who complete gigs as 'gig workers' in the first instance, and as 'app-workers' where we refer to those workers who specifically engage in the app-work variant of gig work.
- We refer to the digital labour platform organisations, which enable and intermediate gig work, as 'platform firms'.
- Finally, we refer to individual consumers who request the services offered by platform firms as 'customers'.

The chapter commences by situating gig working arrangements within the established understanding of non-standard employment, thereby considering the various challenges posed by the 'independent contractor' classification assigned to gig workers. Following this, we focus specifically on the app-work

---

[1]   Funding acknowledgement: this research is funded by the Irish Research Council through the Government of Ireland Postgraduate Scholarship Programme [GOIPG/2018/2196].

variant (that is, app-based gig work which is executed in local markets, such as ride-sharing and food-delivery services) to consider the critical role of technology and the algorithmic management function in facilitating and shaping the working arrangement. Building on these insights, we explore the theoretical issues posed by the disruptive notion of multi-party working relationships in app-work, and the subsequent implications for our traditional conceptualisations of the employment relationship. Drawing on psychological contract theory, we critically examine the individualised nature of these multi-party working relationships, where the governing role of technology shapes the dynamics of the arrangement. Practically, we explore the fragmented, atypical nature of app-working relationships by assessing the interplay and exchanges between each party in the arrangement. In doing so, we contribute to understanding of the social problems related to gig work (for example, precarity and questionable working conditions), explore potential solutions to these issues, and consider the practical and theoretical implications of this heavily digitalised working relationship.

The Nature of Work in the Gig Economy

The dynamic nature of work has left employment relationships in flux. Characterised by changing workplace trends and a fast-paced business environment, recent years have seen a significant shift in how organisations operate (Barley et al. 2017). A combination of factors has contributed to this shift, such as increasing demand for flexibility from workers, a rise in convenience-led consumer demand, and organisational efforts to strategically reduce costs and increase profitability. As a consequence of these and other changes, work itself is likely to become increasingly fragmented as many organisations continue to utilise non-core workers to fill roles traditionally held by salaried employees (Lemmon et al. 2016). These changing workplace trends are especially evident in the gig economy: work is heavily fragmented and issued on a task-by-task basis; workers are classified as independent contractors not employees; roles are advertised as offering high levels of worker flexibility and autonomy; and markets are heavily shaped and often controlled by consumer demand (Kuhn and Maleki 2017; Meijerink and Keegan 2019).

The gig economy, which has developed over the past decade, departs greatly from the once-standard employment model, characterised by full-time, permanent contracts with stable salaries and career progression opportunities (Harvey et al. 2017). Work in the gig economy tends to be built upon radically different foundations when compared with traditional employment relationships (which exists between an employer and employee). For example, as self-employed independent contractors, gig workers are supposedly granted significantly high levels of flexibility in choosing when to work and which tasks to accept. Similarly, platform firms benefit from the flexibility of this working arrangement by avoiding the various financial obligations that typically accompany the legal employment relationship. Yet, these firms simultaneously act as a type of digital intermediary, closely monitoring various aspects of the service-delivery and engagements between gig workers and customers (Duggan et al. 2020). Thus, the gig working relationship appears to consist of a highly defined, narrow and finite set of obligations (Lemmon et al. 2016), that is, a heavily transactional type of arrangement between a service-providing independent contractor and an intermediary platform firm.

If this conceptualisation held true, it would be relatively simple to make sense of working relationships in this new economy based on our existing understanding of various types of contingent labour and the overall nature of transactional work arrangements. However, a multitude of issues, debates and controversies have accompanied discourse on gig working. This includes, among others, concerns about worker exploitation, the erosion of employment standards, power imbalances, questionable working conditions and various legal complexities (Friedman 2014; Stewart and Stanford 2017; Wood et al. 2019). Similarly, attention has focused on the role of technology in gig working, and particularly on the rise of algorithmic management as a novel, alternative tool for monitoring, managing and controlling workers (Duggan et al. 2020; Rosenblat 2018). Given the nature of these concerns, it has been argued that there is considerable scope for gig workers to be considered as employees under the control of platform firms, instead of as independent con- tractors (Collier et al. 2017; Fabo et al. 2017), and that gig workers could, by implication, develop a more nuanced working relationship with the platform itself (Sherman and Morley 2020). Thus, considering the impact of the afore- mentioned complexities, it would be simplistic, and perhaps inappropriate, to assume that all working relationships in the gig economy can be reduced to simple transactional, economic exchanges between independent parties.

MAIN SOCIAL/PRACTICAL PROBLEM

The obfuscation surrounding working relationships in the gig economy is probably attributed to the overall lack of knowledge on what exactly gig work involves, who participates, where they participate and why they participate. In this regard, despite increased focus over the last five years, research on gig working remains in its infancy. Bergman and Jean (2016) note that this may be a consequence of the methodological complexities associated with defining and conceptualising the exact role and motivations of workers in the gig economy, particularly as platform firms and the range of services offered continue to expand and develop. Accordingly, policy-makers and scholars have turned their attention to the gig economy in efforts to better understand how it potentially disrupts the nature of work and impacts on those who participate (Graham et al. 2017). For example, despite the independent contractor classification and outwards similarities with non-traditional forms of work, several key differences exist in gig work that warrant specific consideration (for example, the involvement of multiple parties, and the governing influence and reliance upon technology and digital platforms). Given the lack of homogeneity in gig work, in relation to services offered and platform-based differences, there is no single universal work classification or set of rules that can be implemented across work in the gig economy. Accordingly, many have called for the development of a robust legal architecture characterised by appropriate policy and legislative solutions (Schmidt 2017; Todolí-Signes 2017). Yet, beyond, and perhaps as a precursor to, this discussion, there are critical issues around the dynamic nature of working relationships that require further consideration in the first instance.

Situating Gig Work in Non-standard Employment

In order to make sense of gig working relationships, it is important to first contextualise gig work within the domain of contingent labour. Broadly, working arrangements are becoming increasingly precarious, with significant sectors of the labour market commonly associated with contingent working arrangements and various forms of self-employment or subcontracting (Bonet et al. 2013; Harvey et al. 2017). For example, estimates indicate the prevalence and wide-ranging adaption of non-standard working arrangements, and conditions may, in some capacity, affect up to 75 percent of the global workforce (Cappelli and Kelle, 2013; Harvey et al. 2017). Strategic outcomes of these developments for organisations include a diversified, dynamic workforce without the costs associated with full-time, traditional employment (Harvey et al. 2017). For individuals, the outcomes are more varied. Key advantages may include increased flexibility over where and when to work, diversified income sources, and the opportunity to select projects that best align with an individual's goals and competencies (Weil 2017). Yet, for others, the decline in traditional working arrangements holds greater risk and precarity, characterised by a notable lack of protections, security and career development opportunities (Lemmon et al. 2016).

We propose that gig working arrangements are best described as a hybrid of several contingent work types (Duggan et al. 2020). For example, as per the classification assigned to gig workers, it has similarities to independent contracting (Carr et al. 2017) and forms of subcontracting with the involvement of at least three parties. Gig work also bears similarities to temporary employment, which is neither full time nor open ended (Friedman 2014); temporary agency work, which distributes work via third-party labour intermediaries

(Ward et al. 2001); and zero-hour contracts, where no guaranteed hours are offered (O'Sullivan et al. 2015). Gig working arrangements are undoubtedly characterised by precarity, with a lack of commitment to long-term relationships, flexible working hours, project-based work and piece-rate payments. Yet, for the reasons outlined previously, gig work may prove to be beneficial or desirable for workers with specialised competencies, or for those who wish to earn supplemental income.

Practically, gig work adopts many of the most precarious characteristics and flexible features of existing types of contingent labour. Thus, gig work is both peripheral and perpetual: workers are onboarded quickly into flexible roles, without regard for their past employment, and with no promise for future employment, legacy pay or deferred compensation (Friedman 2014). Gig work introduces a new type of hyper-flexibility to an economy that is already heavily populated with alternative working arrangements, ranging from long-term to short-term, and from formal to informal labour (Van Doorn 2017). Further highlighting the peripheral nature of gig work, platform firms have adopted various names for worker classification in order to distance themselves from any legal employment responsibility (for example, 'taskers' with TaskRabbit and 'riders' with Deliveroo). By hiring independent contractors instead of employees, platform firms can adjust working terms, conditions and payments in response to demand, essentially shifting much of the associated risks on to individual workers (Friedman 2014). The independent worker classification has seen multiple cases being taken by gig workers, with the support of trade unions, challenging the legal status of their employment (McGaughey 2018).

The Role and Influence of Technology

Our consideration of gig work as a hybrid of various forms of contingent labour signals towards the development of a highly transactional, economic type of working relationship. Building on this idea, perhaps the most novel, yet controversial, component of gig work is its reliance on the digital, on-demand features that facilitate the execution of labour. The nature of platform firms, and a critical distinguishing factor of the gig working relationship, is that these firms represent a new way of digitally organising work and offering services. That is, platform firms function as online businesses that digitally facilitate commercial interactions between at least two parties: workers and customers (Gramano 2019). Therefore, it is the digital platform firm that enables the meeting between the gig worker and customer, and, in doing so, mediates this relationship (Gandini 2018). Specifically, in app-work, the scenario is similar: work activities in local markets are conducted through apps, and managed by intermediary digital platform firms that intervene in setting minimum quality standards of service and in the selection and management of individuals who perform the work. These management activities and practices are primarily implemented via algorithms; that is, algorithmic management.

Defined as a system of control where self-learning algorithms are given the responsibility for making and executing decisions affecting labour, the use of algorithmic management in supervising and coordinating gig work heavily impacts upon and complicates the working relationship (Gandini 2018). The technologies utilised provide access to an extremely scalable workforce, eroding long-term relationships and commitment between parties, thereby granting a level of flexibility previously unheard of for the businesses involved (Harvey et al. 2017). Workers are provided just in time and usually compensated on a pay-as-you-go basis; in practice, they are only paid during the moments they actually work (De Stefano 2016), a key distinction from most forms of labour. All communications, instructions and interactions issued to workers only occur virtually, serving to significantly limit the level of human involvement and oversight of labour processes (Veen et al. 2020). In these instances, it can be argued that app-workers subsequently hold a relationship with the governing algorithms embedded in platforms, in place of a relationship with a human representative of the platform firm. This engagement and reliance on a non-human agent may contribute to the creation of a new cohort of invisible workers, operating in isolation or a geographically dispersed manner, under the instruction and guidance of technology-enabled mechanisms (Bankins and Formosa 2020; De Stefano 2016).

MAIN THEORETICAL PROBLEM

As established, the emergence of gig work (and app-work, more specifically) challenges our most basic understanding of the employment relationship (Meijerink and Keegan 2019; Taylor et al. 2017). With the increasingly central role of technology and the independent contractor status assigned to app-workers, the traditional concept of a legal employment relationship between an employer and employee is less applicable. The basis of the gig economy is established upon platform firms supplying branded service offerings to customers, without employing the providers of these services or owning the assets used in the service provision (Sundararajan 2014). Although most app-workers are unlikely to ever actually meet their boss (Barley et al. 2017), the speed at which the working relationship is created is strikingly unique. Typically, workers are onboarded almost instantly via an online application and rapid screening process. Upon commencing roles, workers are generally required to agree to some form of supplier agreement or terms and conditions (that is, in place of a legal contract of employment). With some platforms, workers are required to electronically accept the firm's terms each time they log-in to pursue work opportunities or gigs (Tran and Sokas 2017). App-work participants may therefore, over time, develop a working relationship with the platform firm, owing to the agentic relationship being more transparent.

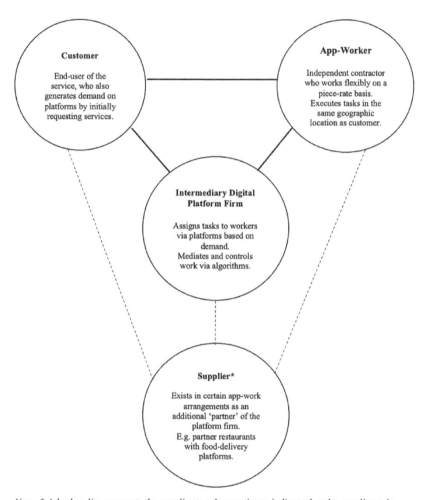

Note: * A broken line connects the supplier to other parties to indicate that the supplier only exists in specific app-working arrangements.
Source: Duggan et al. (2020).

Figure 8.1   Parties involved in app-working relationships

Furthermore, app-work complicates our understanding of the working relationship, with the involvement of at least three parties in the exchange agreement: the platform firm, the app-worker and the customer. Additional parties may also exist in the form of suppliers, who partner with platform firms in the working arrangement. Figure 8.1 illustrates the parties involved in app-working relationships. Interdependencies exist between each of these parties (Meijerink and Keegan 2019), thereby forming a complex working relationship where the worker often holds the least power (Duggan et al. 2020). From the platform firm's perspective, advanced algorithms manage and control exchanges on platforms, ranging from the assignment of tasks, to price-setting, to performance management following the completion of tasks (Van Doorn 2017). Customers also hold a significant role in the working relationship by generating service demand initially, but also in a more crucial capacity by anonymously issuing performance ratings to individual workers (Meijerink and Keegan 2019). Accordingly, instead of simply partnering with platform firms to offer services, most app-workers are at the mercy of digital intermediaries, algorithmic mechanisms and end-user customers.

The Digitalised, Multi-party Working Relationship

Making sense of the multi-party working relationship that exists in app-work is a complex task. This is primarily because the existing legal architecture, seemingly outdated in addressing the unique digital infrastructure inherent in the gig economy, has affected the provision of social and employment protections for app-workers. Therefore, standard employment definitions within these provisions typically exclude those engaged in gig working arrangements (Forde et al. 2017). Here, we adopt an alternative perspective used in understanding the employment relationship, the psychological contract, to help in developing a more cohesive understanding of the individualised nature of app-work (Conway and Briner 2005). Defined as the 'individual beliefs, shaped by the organisation, regarding terms of an exchange agreement between individuals and their organisation', the psychological contract examines the mutual promise-based expectations that parties have of one another and how these implicit expectations impact behaviour (Rousseau 1995, p. 9). It should be noted that the psychological contract differs from the legal contract of employment, as the psychological contract is largely informal, and implicitly rather than explicitly understood (Makin et al. 1996). For example, George (2009) notes that if we were to ask any group of employees what they expect to receive from their organisation and what their organisation expects from them, it is likely that they will have no trouble eliciting a list of terms which would not be found in a conventional contract of employment. By considering app-work from a psychological contract perspective, we seek to develop novel insights into the more tacit dimensions and intricacies of the arrangement. At this period in research on gig working, building knowledge on the process through which individuals establish a working relationship and how the labour arrangement functions, can help us to understand how workers are likely to behave and interact with the other parties in the exchange. Relatedly, by expanding the boundary conditions attached to psychological contract theory through considering how a psychological contract may be created with a non-human agent, such as an algorithm, we make the case that the theory itself has considerable explanatory power in understanding app-working relationships.

While the legal employment contract shapes the psychological contract (Coyle-Shapiro and Kessler 2000), we argue that as many app-workers are currently afforded minimal legal protection this is inconsequential, since most researchers agree that the psychological contract is entirely subjective, with a belief in the existence of an employment relationship at its core (Griep et al. 2018; Rousseau 1989). While psychological contract theory has traditionally been used to help explain the bilateral working agreement between employees and organisations (Rousseau 1995), recent research calls for a multi-foci perspective to better understand the contemporary workplace, where dependencies exist across multiple parties (Alcover et al. 2017). This new psychological contract is symbolic of many workplace trends, such as the rise of alternative arrangements and fragmented work (Bankins et al. 2020), the growth of online labour platforms (Wood et al. 2019) and the development of employment relationships connecting multiple parties (Alcover et al. 2017). Assessing this new, emerging psychological contract will assist theorisation on gig working, but may also assist all contributing parties in addressing the enduring challenge of maintaining effective working relationships beyond solely transactional terms.

Traditionally, when a psychological contract is perceived to exist between an individual and an organisation, the individual implicitly holds expectations of what their employer is obliged to provide to them and what they must contribute in return. These obligations are referred to as the content of the psychological contract (Herriot et al. 1997; Rousseau and Tijoriwala 1999). In comparison to a legal employment contract, there are potentially an infinite number of content dimensions in a psychological contract, given its idiosyncratic nature. Many studies categorise these content dimensions into: (1) relational, reflecting a more open-ended, broader employment relationship, such as developmental opportunities, extra-role behaviours and loyalty; and (2) transactional, reflecting a more closed-ended, narrow, economic employment relationship, such as overtime pay, benefits and working hours (Lam and de Campos 2015; Sherman and Morley 2018). While the app-working relationship is established on heavily transactional terms, emerging research intimates that work in this space may be more idiosyncratic than simply the opportunity to earn an income quickly (Graham et al. 2017; Petriglieri et al. 2018). The value of a psychological contract lens is that it provides insight into the more textured underpinnings of work, particularly in labour arrangements that may, seemingly at least, be largely transactional and routine in focus. In addition, working relationships in the gig economy are notably dynamic, therefore we can expect these to evolve over time with pressures being applied by trade unions, legal challenges, and so on.

Psychological contract fulfilment generally creates feelings of being valued, increases trust, and leads to positive outcomes at an individual and organisational level (Schurer et al. 2003). However, arising from the subjectivity of the psychological contract, breaches and subsequent feelings of violation are common and have serious implications for all parties, including increased exit and decreased loyalty (Turnley and Feldman 1999). Arising when one party perceives a discrepancy between what was received versus what was promised, the psychological contract often becomes truly important only when it is broken in some way, due to the potentially detrimental consequences that accompany breaches (that is, reduced job satisfaction; increased turnover) (Conway and Briner 2005). In the gig economy, the successful functioning of the working arrangement is dependent on the active contribution of multiple parties, all impacting on the relationship in different ways. As a result, psychological contract fulfilment or breach can stem from multiple parties (Alcover et al. 2017). This distributed arrangement, seen as disruptive to the traditional dyadic measures of the theory, further strengthen the argument for a new psychological contract to fully capture the nuances of contemporary labour arrangements (Baruch and Rousseau 2019).

Making Sense of Multi-party App-working Relationships

With multiple, unique sources of dependence across at least three parties, app-work constitutes a working relationship in which platform firms and customers simultaneously generate dependence and determine the rules that shape, afford and limit worker agency (Wood and Lehdonvirta 2019). The concept of agency suggests that individuals express independence through self-assertion and control over environments, work processes or outcomes (Marshall 1989). In psychological contract research, the agency issue has been the subject of long-standing debate (Bankins 2015; Schein 1980), with discussion centred on how individuals may perceive multiple agents as influencing or impacting on roles. Baruch and Rousseau (2019) identified the presence of multiple contract-makers as a key principle of psychological contracts. That is, the psychological contract of any individual worker is shaped by multiple information sources, such as, top management, human resource representatives, career agents and a worker's immediate supervisor. This poses additional challenges for the psychological contract in app-work, where multiple parties shape the arrangement, but where platform firms typically lack human agents. So, with whom does an app-worker hold a psychological contract, and what will be the nature of such a complex arrangement?

We now consider the role of each party in the working arrangement – the app-worker, the platform firm and the customer – to build understanding and unpack the multi-party relationship that exists in app-work. In doing so, we explore how each party is likely to shape the working relationship for individual app-workers through the formation of various interdependencies and power asymmetries.

The app-worker
The psychological contract emerges when one party believes that a promise of future return has been made (for example, pay for performance), a contribution has been given (for example, some form of exchange) and, thus, an obligation has been created to provide future benefits (Robinson and Rousseau 1994; Sturges et al. 2005). While agency is low for app-workers compared with those in traditional employment, we propose that app-work participants may still develop a working relationship (and, subsequently, a psychological contract) with platform firms owing to an agentic relationship being transparent. This is because, in addition to the agreement to terms, workers are required to carry out synchronous work tasks for a local, visible client and remuneration is typically predetermined (Kuhn and Maleki 2017).

Moreover, while app-work relationships are generally not rooted in traditional employee–employer dyads, the involvement of multiple parties interacting and contributing to the dynamic exchange agreement further complicates this (Healy et al. 2017). Trilateral, and sometimes polyadic, relationships exist in app-working arrangements. For example, we consider the case of food-delivery app-work and how agency may be attributed to each party in this arrangement. Across platforms such as UberEats, Deliveroo and Foodora, workers are faced with a triangular working arrangement whereby they serve multiple agents: they act on behalf of the platform firm, but also on behalf of the food outlet or restaurant for whom they are delivering orders (Cochrane and McKeown 2015). These workers are also the only party with direct engagement with the customer who ordered the food, who then has the opportunity to anonymously rate the worker's performance and/or their overall satisfaction with the service.

For these reasons, app-workers are unlikely to identify one specific agent in the working relationship. This multi-dependency indicates that psychological contracts are not simply based on an agreement between two parties, but derived from the multiple actors in the employment network. If, in the above example, the restaurant supplied the worker with the incorrect order, they would be confronted with an unhappy customer upon delivery. The individual worker may perceive this error and subsequent aggravation from the customer as a violation of their psychological contract. Similarly, the worker's performance metrics, generated via algorithmic processes, may be unjustly impacted if this incident results in a delay or complaint, potentially resulting in the untimely exit of a worker. Given the interdependencies between the multiple parties, there are various examples of how such a breach may occur. Thus, the agency issue is fundamental to understanding the perceptions of each party and how the terms of the working arrangement may differ significantly from each individual's viewpoint (Rousseau 1995).

It also becomes apparent that the typical dimensions of employment characteristics, such as negotiation and trust, do not seem to apply in the gig economy (Makoff 2017). This aligns with research suggesting that the realities experienced by many gig workers are significantly different to that which was advertised (Harris 2017, Wood et al. 2019). A key feature of the psychological contract is that the individual voluntarily assents to make and accept specific promises as they understand them. Therefore, it is what the individual believes that they have agreed to which makes the contract, rather than what may have been intended (Rousseau 1995). For example, recent studies illustrate the dissatisfaction and unmet expectations felt by gig workers in relation to power asymmetries, autonomous working and compensation levels (Möhlmann and Zalmanson 2017; Shapiro 2018).

Yet, as the app-worker may perceive dissatisfaction and unmet expectations from different agents (that is, perhaps from the platform firm in one instance and from the restaurant in another), this raises the possibility of multiple psychological contracts existing within a single working relationship (Wiechers et al. 2019). To add to this complexity, we must also consider whether the app-worker may encounter a heavily transactional relationship with one party, while simultaneously holding a more relational arrangement with another. With the norm of reciprocity being unclear in complex multi-party arrangements, this can be problematic if one party feels that a breach has taken place (Baruch and Rousseau 2019). That is, we should not assume that all app-working arrangements are entirely transactional, simply because the relationship has been established as such by the platform firm.

The platform firm
One of the most challenging aspects of app-working relationships is that one of the parties in the arrangement is the app itself, which, as explained, is controlled by the platform firm and utilises algorithmic management. Platform firms both maintain and distance themselves from responsibility over their markets: on the one hand, they retain control over the allocation of work, conditions and minimum performance standards; on the other, they deny many responsibilities by identifying only as technological companies who serve to provide a medium of exchange (Healy et al. 2017). A key feature of app-work is the automatic coordination and matching of the transaction through a set of advanced algorithms, creating a space where supply and demand inte-

grate through automatic management and enforced mechanisms (Lehdonvirta 2018). That is, platform firms use algorithms to match supply and demand in the market, while also mediating and closely monitoring the work performed (Gandini 2018). In view of this, when discussing the role of the platform firm, we most commonly refer to either the app or the algorithmic management function.

As a technology-enabled monitoring tool, algorithmic management eliminates the more interpersonal and empathetic aspects of people management. By automating management practice, algorithms exert control over workforces and facilitate asymmetrical information in the working relationship. In app-work, this takes many forms, including: controlling the supply of labour; targeting workers with incentives; removing workers from platforms; and mediating disputes at its own discretion (Duggan et al. 2020). Also, many platform firms fully exploit the possibility to minutely monitor the activities of app-workers in real time, rendering them heavily at the mercy of the platform's expectations and demands (Kuhn and Maleki 2017). This always-on type of control is seen as unfair by some app-workers, particularly given their legal employment classification (Gandini 2018). Without a human agent advocating their needs on behalf of the platform firm to deliver a more balanced working relationship (Gilbert et al. 2011), app-workers may lose trust and confidence, resulting in a reduced sense of well-being.

Yet, research tells us that firms cannot hold psychological contracts, but they provide context for the creation of the psychological contract and may instead be represented by human agents of the organisation (that is, managerial representatives) (Guest 1998; Rousseau 1989). With the lack of human organisational representatives in app-work, the agency question becomes especially apposite. Is it possible that an app, governed by algorithmic management mechanisms, be considered a party in the development of a psychological contract? In its role as intermediary, the platform firm is the only party with full access to, and control over, the data, processes and rules of the platform (Jabagi et al. 2019). Ravenelle (2019) argues that the capricious nature of apps may impact workers' experiences of psychological contract violation or fulfilment. As noted by Jago (2019), people believe that technological agents lack the same level of moral authenticity as human agents. This arm's-length approach to managing workers, characterised by data-driven performance statistics and metrics, signals the development of a heavily transactional psychological contract. However, research by Bankins and Formosa (2020) argues that through increasing sophistication, artificially intelligent technologies are now moving away from being viewed as a simple tool used by organisations, towards being seen as partners of and within organisations. Findings such as this support the argument that the app-worker, in part at least, may develop more relational expectations of the app itself, even if the platform firm rejects this notion.

The customer

The third party in app-working relationships is the customer. Although ostensibly the most passive party in the working relationship, customers potentially play an influential role in shaping the psychological contracts of app-workers as many platform firms utilise customer ratings of workers via anonymous systems as a means of performance evaluation. Before being shared with app-workers, assessments are forwarded to platform firms, who have the opportunity to verify, albeit indirectly, the quality and punctuality of the service rendered (Gramano 2019). These data, along with customer ratings and reviews, may then be used to identify the best performing workers and to alter the algorithm that assigns tasks to workers. In view of this, most app-workers are subject to tight levels of control, where the customer evaluation is often critical (Healy et al. 2017), and without recourse to question or refute performance scores. In some instances, the rating issued by the customer may not solely reflect the individual worker's performance, but the overall service delivery. This can potentially be detrimental for workers. Again, in the case of food-delivery app-work, although customer orders and requests are sent to restaurants via the platform firm, the app-worker is the only party with direct, human interaction with the customer. As a consequence, the worker, through the receipt of performance ratings, is at most risk of being penalised for errors potentially made by other parties in the working arrangement (Duggan et al. 2020; Goods et al. 2019). This arguably results in a generally one-sided process (Flanagan 2018), which intimates a further shift in power away from the app-worker.

As app-workers are potentially rated after the completion of each individual task, the combined roles of the customer and the platform firm essentially exercise penetrating control over all aspects of the service delivery. The worker can be described as being in a situation comparable to a permanent probationary period, again signalling a more transactional arrangement (Gramano 2019). Yet, given the fundamental role of customer ratings, in that workers in receipt of low ratings are potentially subjected to being deactivated (Tran and Sokas 2017), concerns are raised about the inherent weaknesses in simplistic and entirely quantitative measures of performance (McDonnell et al. 2019). For example, in some cities, Uber drivers with an average rating, calculated by an algorithm, of lower than 4.6 out of a possible 5 are at risk of disbarment from the platform (Kuhn and Maleki 2017). Similarly, considering the customer's role, another risk is that app-work is supplied entirely through information technology channels, whether these are online platforms or apps, thereby potentially distorting the perceptions customers may have of these workers and significantly contributing to a perceived dehumanisation of their activity (De Stefano 2016).

CONSEQUENCES OF THE SOCIAL/PRACTICAL PROBLEM

While it is important, in the context of the psychological contract, to understand the role of each party in app-work arrangements in isolation, it is more pressing from a practical perspective to explore the interplay and exchanges between each party. It is the dynamic interchange between parties through which the psychological contract develops (Conway and Briner 2005). For instance, Sherman and Morley (2020) suggest that for app-workers, the psychological contract held with one party impinges on the contract created with the other party, resulting in complications that arise owing to multiple dependence. Specifically, they found that app-workers in food-delivery arrangements expressed frustration that restaurant delays impact upon their relationship with the customer. Findings such as this indicate a sense of the emotional resources that an app-worker has to expend to fulfil an obligation. This is a significant issue in multi-party working arrangements as there may be consequences for other parties if an individual has depleted resources after expending a lot of energy in trying to fulfil an obligation for one particular party (Deng et al. 2018; Sherman and Morley 2020). In addition, the nature of multi-party working relationships may mean that a breach of the psychological contract with one party can be triggered by the relationship with another, and, perhaps, that the fulfilment of one agreement may come at the expense of another (Wiechers et al. 2019). These studies illustrate the complexities at the heart of multi-party working arrangements, and raise questions about how these relationships can be effectively managed.

A heavily transactional relationship appears to be at the heart of app-work, with workers paid for the quantity of work undertaken rather than the time spent working. The seemingly non-existent focus on the development of mutual trust and commitment in the working relationship further solidifies the transactional nature of this exchange. For example, at the recruitment stage, most roles are advertised on the basis that app-workers have the autonomy to work when they wish, with considerable independence, indicating that there are little or no expectations of a long-term relationship (Jabagi et al. 2019). Workers are typically onboarded quickly, via a prompt screening process, ensuring a readily accessible source of labour for the platform firm (Kuhn and Maleki 2017). While this approach to onboarding workers certainly reduces costs by eliminating many of the labour and time costs involved (Healy et al. 2017), it may prove problematic when looking beyond short-term, transactional cost benefits towards recruiting motivated workers who are likely to succeed in their roles.

While the transactional aspect of app-work is evident, we argue that this conceptualisation may under-appreciate the inherently nuanced structure of the arrangement, which forms a working relationship with a minimum of three parties involved (Wood and Lehdonvirta 2019). For example, emerging evidence signals that app-workers are seeking a more co-determined and relational work arrangement, populated by opportunities to develop new skills useful in furthering their careers, in addition to craving social interaction and networking opportunities (Maffie 2020; Petriglieri et al. 2018). Seeking greater voice in the working relationship and opportunities for professional growth are typical content dimensions of a relational psychological contract (Conway and Briner 2005). So, from this perspective, there appears to be more to app-work than simply remuneration and flexibility. It seems that, simultaneously, app-work and its use of algorithmic management both enables and contorts the interactions between each party in the arrangement.

Consequently, psychological contract theory may be useful in this context, as the perception of each party is of critical importance and the terms may differ significantly based on these perceptions (Rousseau 1995). This theory appreciates that one party can perceive a particular relationship and expectations that the other does not recognise. Thus, while a legal employment contract may be denied by the platform firm, it is feasible that psychological contracts exist and develop, at least from the worker's perspective. The idea that app-workers only seek to work for financial gain appears at odds with recent research and wider scholarship on the sociology of work, around the relevance of relational arrangements (Woodcock and Johnson 2018; Wood and Lehdonvirta 2019).

SOLUTIONS TO THE SOCIAL/PRACTICAL PROBLEM

The complexities of app-work, combined with the subjective nature of the psychological contract, make it difficult to propose solutions to how this fragmented working relationship can be effectively managed. Perhaps the most obvious and widely discussed path to resolving the issues inherent in gig work is a legal one; that is, revising or updating existing employment legislation to effectively govern the gig economy. It is unsurprising, then, that a great deal of scholarly literature and broader discourse on gig work has considered, at least to some extent, potential remedies to fix legal and policy architectures in this domain (Bonet et al. 2013; Tran and Sokas 2017). These discussions have focused on issues such as labour exploitation, taxation, and worker classification (Prassl 2018; Schiek and Gideon 2018). A commonly explored solution is the proposal to create a special employment status for gig workers, acknowledging that the characteristics of these individuals do not fully comply with either the definition of employee or the self-employed (Maselli et al. 2016). The rationale is the creation of flexicurity for workers, signalling a type of intermediate classification that acknowledges the flexibility of gig work, while also providing some limited form of labour protection (De Stefano 2016). Ostensibly, the benefits brought about by this change may well be widespread: this solution potentially fills the regulatory gap affecting the gig economy, while also managing expectations by more clearly defining the specific nature of the working relationship.

However, the process of implementing a solution of this nature is far from straightforward. For example, recent years have seen a multitude of different rulings on the same issues regarding gig workers, with similar types of app-workers being determined as employees or independent contractors across different jurisdictions (Johnston and Land-Kazlauskas 2018; McGaughey 2018). At the time of writing, the saga is ongoing, with the latest developments on California's Assembly Bill No. 5 (AB5), which, since being passed in 2019, extends employee classification to gig workers (Schor et al. 2020). However, California's gig economy fought back by getting Proposition 22 – the 'App-based drivers as contractors and labour policies initiative' – accepted, which essentially grants platform firms an exemption from AB5, freeing them from complying with California's labour laws (Farrell 2020). It is becoming increasingly apparent that many larger platform firms will persist in contesting efforts to regulate the gig economy, making it unlikely that a universal solution of this nature will be accomplished with any ease.

In view of this, in the interim, we propose that it is beneficial to concentrate elsewhere by seeking to resolve the specific issues that arise between parties in app-work. As illustrated throughout this chapter, the relationships between the multiple parties in the working arrangement potentially result in the formation of several psychological contracts, all of which vary in nature and depth. Identifying and addressing the most pertinent imbalances in each of these relationships is likely to provide a useful foundation in the formation of a functional working arrangement. Although empirical research is required to definitively uncover these issues and imbalances, we briefly consider areas where issues seem likely to arise from the app-worker's perspective.

The social and relational aspects of app-work are heavily shaped by technology and digitalised management tools (Duggan et al. 2020). Although the use of algorithmic management enables increased efficiency and innovation, the lack of human representation and engagement on behalf of the platform firm weakens social ties and forms perhaps the largest obstacle for individuals seeking to craft a relational working arrangement (Wang et al. 2020). However, for those seeking a more casual arrangement, characterised by increased flexibility and liberation from the confines of a traditional office environment, the heavily digitalised nature of the app-working relationship may be more desirable. Thus, as is the subjective nature of the psychological contract, issues are most likely to occur based on an individual app-worker's expectations of the role, and whether these expectations are aligned with what is offered by the platform firm (Sherman and Morley 2020). While this may be so in any role, the blurring of lines in app-work highlights these issues. Accordingly, a key starting-point in solving this issue lies in effective, accurate communications to appropriately manage the expectations of app-workers at the recruitment and onboarding stages, making clear what the working relationship entails and, importantly, what it does not entail.

Relatedly, beyond the core relationship between the app-worker and the platform firm, issues may also arise from workers' engagements with other parties in the arrangement. For example, the role of customers in app-work is complex and warrants consideration. For food-delivery app-workers, the relationship formed with customers is likely to be fleeting and heavily trans-actional, relying almost entirely on the timely delivery of an order (Duggan et al. 2020). In contrast, a ride-share app-worker's relationship with customers is likely to be significantly more nuanced, lasting for the entire duration of the journey, and heavily shaped by each customer's preferences for interpersonal engagement and interaction (Rosenblat 2018). In view of this, perhaps a poten-tial solution for app-workers may be to move towards informally negotiating their performances with customers, thereby developing more relational strate-gies within this relationship (Vallas and Schor 2020).

Finally, we are also concerned with emerging evidence that app-workers seek to develop competencies useful in furthering their careers, and the potential lack of such opportunities within existing arrangements (Maffie 2020; Petriglieri et al. 2018). Thus, the issue to be addressed in this regard lies in striking the balance between flexibility, sustainability and opportunity for app-workers. Some platform firms have already engaged in this debate by offering potential solutions. For example, Uber recently launched 'Uber Works', a new extension of its service that connects casual workers with businesses offering short-term roles in various sectors, allowing individuals to compare pay rates and sign up for shifts that suit their schedules (Uber 2019). The Deliveroo Rider Academy offers workers access to online learning courses, apprenticeships and mentoring opportunities to develop their careers. However, although these initiatives are an encouraging starting point, the broader issue regarding the potential unsustainability of app-work remains (Ashford et al. 2018).

DIRECTIONS FOR FUTURE RESEARCH

The discourse around the rise of the gig economy has grown exponentially during the past decade. The gig economy provides challenges to established thought regarding the nature of work and the employment relationship. Despite recognising that the gig economy is disruptive to our understanding of employment relationships, specific details surrounding this are lacking, as is an understanding of the subsequent implications as these relationships evolve. In this chapter, we have explored the nature of multi-party working relationships in the gig economy. By focusing explicitly on app-work, we utilised a psychological contract perspective to consider the more tacit, idiosyncratic dimensions of the exchange agreement that exists between the parties in the arrangement. Most platform firms deny the existence of an employment relationship, but the strict levels of control enforced raise noteworthy contradictions that warrant developed empirical consideration (Kuhn and Maleki 2017). Utilising psychological contract theory potentially uncovers new depths to the relationship between app-workers, the platform firms for whom they work and the end-user customers who generate demand. Accordingly, conflict between workers and firms on issues surrounding control, dependency and working conditions, for example, would suggest that there is more to the working relationship than simply remuneration and flexibility.

Thus, app-work disrupts the contours of our traditional understanding of the employment relationship by introducing changes in where, how and what work is completed for firms (Bankins et al. 2020). The nature of this work results in many fundamental aspects of a more traditional, transparent employment relationship being enabled and delivered exclusively via digital platforms (Kuhn and Maleki 2017). Owing to this, the evidence of an agentic relationship between the multiple parties participating in app-work suggests that workers may still develop a psychological contract with the platform firms for whom they work. This challenges the long-held belief in psychological contract research that only humans can serve as contract-makers (Griep et al. 2019; Sherman and Morley 2020). The spatial and temporal boundaries of app-work lend further weight to the argument that working relationships in the gig economy are atypical in the business landscape, in that they can both restrict and facilitate the interaction between the multiple parties in the working arrangement. Indeed, that app-work relationships are generally not rooted in traditional employee-employer dyads but instead involve multiple parties contributing to the dynamic exchange agreement, presents challenges to how the working relationship can be managed effectively.

Thus, empirical research is warranted to thoroughly understand the impact of these relatively novel and alternative digitalised management practices for workers, and indeed for the people management function as the workplace continues to change (Bankins and Formosa 2020). More broadly, perhaps, this is particularly relevant currently as the world grapples with and emerges from the COVID-19 pandemic. The societal implications of this global crisis pose significant and unprecedented challenges for the future of work and workplace design, especially where the need for increased remote working has come to the fore. Similarly, gig workers across various industries have, in many cases, been central to the delivery of key services during the pandemic, while simultaneously being one of the most at-risk groups and receiving little by way of social and employment benefits and protections (Griswold 2020).

Shared understandings and reciprocal contributions for mutual benefit are at the core of traditional, functional exchange relationships between workers and organisations (Dabos and Rousseau 2004). From the app-worker's perspective, having no say in how work is assigned and how performance is assessed means that the working relationship is less an arrangement that has been mutually co-determined, and could instead be perceived as a working arrangement of subjugation (Harvey et al. 2017). Also, the fragmented nature of app-work, through its reliance on technology via digital platforms and governing algorithms, may erode the reciprocity found in traditional employment relationships. The unique, alternative working arrangements found on the majority of app-work platforms present significant and novel challenges for scholars and practitioners alike regarding how to effectively manage the employment relationship in this domain. Therefore, we believe that future research pursuing the lines of enquiry outlined in this chapter hold the prospect of providing important insight into this understudied, dynamic area.

REFERENCES

Alcover, C.M., R. Rico, W.H. Turnley and M.C. Bolino (2017), 'Multi-dependence in the formation of the distributed psychological contract', *European Journal of Work and Organisational Psychology*, **26** (1), 16–29.

Ashford, S.J., B.B. Caza and E.M. Reid (2018), 'From surviving to thriving in the gig economy: a research agenda for individuals in the new world of work', *Research in Organizational Behaviour*, **38** (December), 23–41.

Bankins, S. (2015), 'A process perspective on psychological contract change: making sense of, and repairing, psychological contract breach and violation through employee coping actions', *Journal of Organisational Behaviour*, **36** (8), 1071–95.

Bankins, S. and P. Formosa (2020), 'When AI meets PC: exploring the implications of workplace social robots and a human-robot psychological contract', *European Journal of Work and Organizational Psychology*, **29** (2), 215–29.

Bankins, S., Y. Griep and S.D. Hansen (2020), 'Charting directions for a new research era: addressing gaps and advancing scholarship in the study of psychological contracts', *European Journal of Work and Organizational Psychology*, **29** (2), 159–63.

Barley, S.R., B.A. Bechky and F.J. Milliken (2017), 'The changing nature of work: careers, identities, and work lives in the 21st century', *Academy of Management Discoveries*, **3** (2), 111–15.

Baruch, Y. and D.M. Rousseau (2019),' Integrating psychological contracts and ecosystems in career studies and management', *Academy of Management Annals*, **13** (1), 84–111.

Bergman, M.E. and V.A. Jean (2016), 'Where have all the "workers" gone? A critical analysis of the unrepresentativeness of our samples relative to the labour market in the industrial-organisational psychology literature', *Industrial and Organisational Psychology*, **9** (1), 84–113.

Bonet, R., P. Cappelli and M. Hamori (2013), 'Labour market intermediaries and the new paradigm for human resources', *Academy of Management Annals*, **7** (1), 341–92.

Cappelli, P. and J.R. Keller (2013), 'Classifying work in the new economy', *Academy of Management Review*, **38** (4), 575–96.

Carr, C.T., R.D. Hall, A.J. Mason and E.J. Varney (2017), 'Cueing employability in the gig economy: effects of task-relevant and task-irrelevant information on Fiverr', *Management Communication Quarterly*, **31** (3), 409–28.

Cochrane, R. and T. McKeown (2015), 'Vulnerability and agency work: from the workers' perspectives', *International Journal of Manpower*, **36** (6), 947–65.

Collier, R.B., V.B. Dubal and C. Carter (2017),' Labour platforms and gig work: the failure to regulate', Working Paper No. 106–17, Institute for Research on Labour and Employment, Berkeley, CA.

Conway, N. and R.B. Briner (2005), *Understanding Psychological Contracts at Work: A Critical Evaluation of Theory and Research*, New York: Oxford University Press.

Coyle-Shapiro, J. and I. Kessler (2000), 'Consequences of the psychological contract for the employment relationship: a large-scale survey', *Journal of Management Studies*, **37** (7), 903–30.

Dabos, G.E. and D.M. Rousseau (2004), 'Mutuality and reciprocity in the psychological contracts of employees and employers', *Journal of Applied Psychology*, **89** (1), 52–72.

De Stefano, V. (2016), *The Rise of the 'Just-in-Time' Workforce: On-demand Work, Crowdwork and Labour Protection in the Gig Economy*, Geneva: International Labour Office.

Deng, H., J. Coyle-Shapiro and Q. Yang (2018), 'Beyond reciprocity: a conservation of resources view on the effects of psychological contract violation on third parties', *Journal of Applied Psychology*, **103** (5), 561–77.

Duggan, J., U. Sherman, R. Carbery and A. McDonnell (2020), 'Algorithmic management and app-work in the gig economy: a research agenda for employment relations and HRM', *Human Resource Management Journal*, **30** (1), 114–32.

Fabo, B., J. Karanovic and K. Dukova (2017), 'In search of an adequate European policy to the platform economy', *Transfer: European Review of Labour and Research*, **23** (2), 163–75.

Farrell, H. (2020), 'Uber wants to limit its drivers' rights in California. User loyalty is its secret political weapon', *Washington Post*, 21 August, accessed 28 August 2020 at https://www.washingtonpost.com/politics/2020/08/21/uber-wants-limit-its-drivers-rights-california-user-loyalty-is-its-secret-political-weapon.

Flanagan, F. (2018), 'Theorising the gig economy and home-based service work', *Journal of Industrial Relations*, **61** (1), 57–78.

Forde, C., M. Stuart, S. Joyce, L. Oliver, D. Valizade, G. Alberti, et al. (2017), 'The social protection of workers in the platform economy', *Policy Department A: Economic and Scientific Policy*, accessed at http://www.europarl.europa.eu/ RegData/etudes/STUD/2017/614184/IPOL_STU(2017)614184_EN.pdf.

Friedman, G. (2014), 'Workers without employers: shadow corporations and the rise of the gig economy', *Review of Keynesian Economics*, **2** (2), 171–88.

Gandini, A. (2018), 'Labour process theory and the gig economy', *Human Relations*, **72** (6), 1039–56.

George, C. (2009), *The Psychological Contract: Managing and Developing Professional Groups*, Maidenhead: Open University Press.

Gilbert, C., S. De Winne and L. Sels (2011), 'The influence of line managers and HR department on employees' affective commitment', *International Journal of Human Resource Management*, **22** (8), 1618–37.

Goods, C., A. Veen and T. Barratt (2019), '"Is your gig any good?" Analysing job quality in the Australian platform-based food-delivery sector', *Journal of Industrial Relations*, **61** (4), 502–27.

Graham, M., I. Hjorth and V. Lehdonvirta (2017), 'Digital labour and development: impacts of global digital labour platforms and the gig economy on worker liveli-hoods', *Transfer: European Review of Labour and Research*, **23** (2), 135–62.

Gramano, E. (2019), 'Digitalisation and work: challenges from the platform economy', *Contemporary Social Science*, **15** (4), 476–88.

Griep, Y., C. Cooper, S. Robinson, D. Rousseau, S.D. Hansen, M. Tompreu, et al. (2019), 'Psychological contracts: back to the future', in Y. Griep and C. Cooper (eds), *Handbook of Research on the Psychological Contract at Work* , Cheltenham, UK and Northamptopn, MA, USA: Edward Elgar, pp. 397–414.

Griep, Y., T Vantilborgh and S.K. Jones (2018), 'the relationship between psycholog-ical contract breach and counterproductive work behaviour in social enterprises: do paid employees and volunteers differ?', *Economic and Industrial Democracy*, **41** (3), 727–45.

Griswold, A. (2020), 'The month the entire world signed up for delivery', *Quartz*, accessed 28 August 2020 at https://qz.com/1838349/how-coronavirus-will-change -the-online-delivery-business/.

Guest, D.E. (1998), 'Is the psychological contract worth taking seriously?', *Journal of Organisational Behaviour*, **19** (S1), 649–64.

Harris, B. (2017), 'Uber, Lyft, and regulating the sharing economy', *Seattle University Law Review*, **41** (1), 269–85.

Harvey, G., C. Rhodes, S.J. Vachhani and K. Williams (2017), 'Neo-villeiny and the service sector: the case of hyper flexible and precarious work in fitness centres', *Work, Employment and Society*, **31** (1), 19–35.

Healy, J., D. Nicholson and A. Pekarek (2017), 'Should we take the gig economy seriously?', *Labour and Industry: A Journal of the Social and Economic Relations of Work*, **27** (3), 232–48.

Herriot, P., W.E.G. Manning and J.M. Kidd (1997), 'The content of the psychological contract', *British Journal of Management*, **8** (2), 151–62.

Jabagi, N., A.M. Croteau, L.K. Audebrand and J. Marsan (2019), 'Gig workers' moti-vation: thinking beyond carrots and sticks', *Journal of Managerial Psychology*, **34** (4), 192–213.

Jago, A.S. (2019), 'Algorithms and authenticity', *Academy of Management Discoveries*, **5** (1), 38–56.

Johnston, H. and C. Land-Kazlauskas (2018), *Organising On-demand: Representation, Voice, and Collective Bargaining in the Gig Economy*, Geneva: International Labour Office, accessed 28 August 2020 at https://www.ilo.org/wcmsp5/groups/public/---ed_protect/---protrav/---travail/documents/publication/wcms_624286.pdf.

Kuhn, K.M. and A. Maleki (2017), 'Micro-entrepreneurs, dependent contractors, and Instaserfs: understanding online labour platform workforces', *Academy of Management Perspectives*, **31** (3), 183–200.

Lam, A. and A. de Campos (2015), '"Content to be sad" or "runaway apprentice"? The psychological contract and career agency of young scientists in the entrepreneurial university', *Human Relations*, **68** (5), 811–41.

Lehdonvirta, V. (2018), 'Flexibility in the gig economy: managing time on three online piecework platforms', *New Technology, Work and Employment*, **33** (1), 13–29.

Lemmon, G., M.S Wilson, M. Posig and B.C. Glibkowski (2016), 'Psychological contract development, distributive justice, and performance of independent contractors: the role of negotiation behaviours and the fulfilment of resources', *Journal of Leadership and Organisational Studies*, **23** (4), 424–39.

Maffie, M.D. (2020), 'The role of digital communities in organising gig workers', *Industrial Relations: A Journal of Economy and Society*, **59** (1), 123–49.

Makin, P.J., C.L. Cooper and C.J. Cox (1996), *Organisations and the Psychological Contract: Managing People at Work*, Leicester: Wiley-Blackwell.

Makoff, A. (2017), 'Nearly eight million people thinking about joining gig economy, study reveals', *CIPD: People Management*, 27 April, accessed 28 August 2020 at https://www.peoplemanagement.co.uk/news/articles/eight-million-people-thinking-joining-gig-economy.

Marshall, J. (1989), 'Re-visioning career concepts: a feminist invitation', in M.B. Arthur, D.T. Hall and B.S. Lawrence (eds), *Handbook of Career Theory*, New York: Cambridge University Press, pp. 275–91.

Maselli, I., K. Lenaerts and M. Beblavy (2016), 'Five things we need to know about the on-demand economy', Centre for European Policy Studies, Brussels.

McDonnell, A., P. Gunnigle and K.R. Murphy (2019), 'Performance management', in D.G. Collings, G. Wood and L. Samosi (eds), *Human Resource Management: A Critical Approach*, London: Routledge, pp. 189–207.

McGaughey, E. (2018), 'Taylorooism: when network technology meets corporate power', *Industrial Relations Journal*, **49** (5–6), 459–72.

Meijerink, J. and A. Keegan, A. (2019), 'Conceptualizing human resource management in the gig economy: toward a platform ecosystem perspective', *Journal of Managerial Psychology*, **34** (4), 214–32.

Möhlmann, M. and L. Zalmanson (2017), 'Hands on the wheel: navigating algorithmic management and Uber drivers' autonomy', in Y.J. Kim, R. Agarwal and J.K. Lee (eds), *Proceedings of the International Conference on Information Systems: Transforming Society with Digital Innovation, ICIS 2017, Seoul, 10–13 December*, Atlanta, GA: Association for Information Systems.

O'Sullivan, M., T. Turner, J. McMahon, L. Ryan, J. Lavelle, C. Murphy, et al. (2015), 'A study on the prevalence of zero hours contracts among Irish employers and their impact on employees', Kemmy Business School, University of Limerick, accessed 28 August 2020 at https://dbei.gov.ie/en/Publications/Publication-files/Study-on-the-Prevalence-of-Zero-Hours-Contracts.pdf.

Petriglieri, G., S.J. Ashford and A. Wrzesniewski (2018), 'Agony and ecstasy in the gig economy: cultivating holding environments for precarious and personalised work identities', *Administrative Science Quarterly*, **64** (1), 124–70.

Prassl, J. (2018), *Humans as a Service*, Oxford: Oxford University Press.

Ravenelle, A.J. (2019), '"We're Not Uber": control, autonomy, and entrepreneurship in the gig economy', *Journal of Managerial Psychology*, **34** (4), 269–85.

Robinson, S.L. and D.M. Rousseau (1994), 'Violating the psychological contract: not the exception but the norm', *Journal of Organisational Behaviour*, **15** (3), 245–59.

Rosenblat, A. (2018), *Uberland: How Algorithms Are Rewriting the Rules of Work*, Oakland, CA: University of California Press.

Rousseau, D.M. (1989), 'Psychological contract and implied contracts in organisations', *Employee Responsibilities and Rights Journal*, **2** (2), 121–39.

Rousseau, D.M. (1995), *Psychological Contracts in Organisations: Understanding Written and Unwritten Agreements*, London: Sage.

Rousseau, D.M. and S. Tijoriwala (1999), 'Assessing psychological contracts: issues, alternatives and measures', *Journal of Organisational Behaviour*, **84** (4), 514–28.

Schein, E.H. (1980), *Organisational Psychology*, Englewood Cliffs, NJ: Prentice Hall.

Schiek, D. and A. Gideon (2018), 'Outsmarting the gig economy through collective bargaining – EU competition law as a barrier to smart cities', *International Review of Law, Computers and Technology*, **32** (2–3), 275–94.

Schmidt, F.A. (2017), 'Digital labour markets in the platform economy: mapping the political challenges of crowd work and gig work', Friedrich-Ebert-Stiftung, Bonn.

Schor, J.B., W. Attwood-Charles, M. Cansoy, L. Ladegaard and R. Wengronowitz (2020), 'Dependence and precarity in the platform economy', *Theory and Society*, **49** (October), 833–61.

Schurer, L.L., J.R. Edwards and D.M. Cable (2003), 'Breach and fulfilment of the psychological contract: a comparison on traditional and expanded views', *Personnel Psychology*, **56** (4), 895–934.

Shapiro, A. (2018), 'Between autonomy and control: strategies of arbitrage in the on-demand economy', *New Media and Society*, **20** (8), 2954–71.

Sherman, U.P. and M.J. Morley (2018), 'Organizational inputs to the formation of the expatriate psychological contract: towards an episodic understanding', *International Journal of Human Resource Management*, **29** (8), 1513–36.

Sherman, U.P. and M.J. Morley (2020), 'What do we measure and how do we elicit it? The case for the use of repertory grid techniques in multi-party psychological contract research', *European Journal of Work and Organizational Psychology*, **29** (2), 230–42.

Stewart, A. and J. Stanford (2017), 'Regulating work in the gig economy: what are the options?', *Economic and Labour Relations Review*, **28** (3), 420–37.

Sturges, J., N. Conway, D. Guest and A. Liefooghe (2005), 'Managing the career deal: the psychological contract as a framework for understanding career management, organisational commitment and work behaviour', *Journal of Organisational Behaviour*, **26** (7), 821–38.

Sundararajan, A. (2014), 'What Airbnb gets about culture that Uber doesn't', *Harvard Business Review*, 27 November, accessed 28 August 2020 at https://hbr.org/2014/11/what-airbnb-gets-about-culture-that-uber-doesn't.

Taylor, M., G. Marsh, D. Nicol and P. Broadbent (2017), *Good Work: The Taylor Review of Modern Working Practices*, London: Department of Business, Energy and Industrial Strategy.

Todolí-Signes, A. (2017), 'The gig economy: employee, self-employed or the need for a special employment regulation?', *Transfer: European Review of Labour and Research*, **23** (2), 193–205.

Tran, M. and R.K. Sokas (2017), 'The gig economy and contingent work: an occupational health assessment', *Journal of Occupational and Environmental Medicine*, **59** (4), 63–6.

Turnley, W.H. and D.C. Feldman (1999), 'The impact of psychological contract violations on exit, voice, loyalty, and neglect', *Human Relations*, **52** (7), 895–922.

Uber (2019), 'Uber Works officially launched in Chicago', *Uber*, blog, accessed 28 August 2020 at https://www.uber.com/blog/chicago/uberworks/.

Vallas, S. and J.B. Schor (2020), 'What do platforms do? Understanding the gig economy', *Annual Review of Sociology*, **46**, (July), 273–94.

Van Doorn, N. (2017), 'Platform labor: on the gendered and racialized exploitation of low-income service work in the "on-demand" economy', *Information, Communication & Society*, **20** (6), 898–914.

Veen, A., T. Barratt and C. Goods (2020), 'Platform-capital's "app-etite" for control: a labour process analysis of food-delivery work in Australia', *Work, Employment and Society*, **34** (3), 388–406.

Wang, B., Y. Liu and S.K. Parker (2020), 'How does the use of information communication technology affect individuals: a work design perspective', *Academy of Management Annals*, **14** (2), 695–725.

Ward, K., D. Grimshaw, J. Rubery and H. Beynon (2001), 'Dilemmas in the management of temporary work agency staff', *Human Resource Management Journal*, **11** (4), 3–21.

Weil, D. (2017), 'How to make employment fair in an age of contracting and temp work', *Harvard Business Review*, 24 March, accessed 28 August 2020 at https://hbr.org/2017/03/making-employment-a-fair-deal-in-the-age-of-contracting -subcontracting-and-temp-work.

Wiechers, H., J.A. Coyle-Shapiro, X.D. Lub and S. Ten Have (2019), 'Triggering psychological contract breach', in Y. Griep and C. Cooper (eds), *Handbook of Research on the Psychological Contract at Work*, Cheltenham, UK and Northampton, MA, USA: Edward Elgar, pp. 272–91.

Wood, A.J. and V. Lehdonvirta (2019), 'Platform labour and structured antagonism: understanding the origins of protest in the gig economy', paper presented at the Oxford Internet Institute Platform Economy Seminar Series, 5 March, Oxford.

Wood, A.J., M. Graham, V. Lehdonvirta and I. Hjorth (2019), 'Good gig, bad gig: autonomy and algorithmic control in the global gig economy', *Work, Employment and Society*, **33** (1), 56–75.

Woodcock, J. and M.R. Johnson (2018), 'Gamification: what it is, and how to fight it', *Sociological Review*, **66** (3), 542–58.

PART III

Solutions and conclusions

9. Gigs of their own: reinventing worker cooperativism in the platform economy and its implications for collective action

Damion Jonathan Bunders

MAIN SOCIAL PROBLEM: THE PUZZLE OF COLLECTIVE ACTION

The recent rise of gig platforms brings together longer trends of work flexibilisation, digital technology and changing consumer preferences to act as intermediaries between local on-demand via app workers or remote crowdworkers and their respective customers (De Stefano 2015; Woodcock and Graham 2020). On the one hand, gig platforms may generate flexible job opportunities and new ways to make a living (Martin 2016). Their low entry barrier, and for crowdwork access to global labour markets, makes them particularly attractive to disadvantaged groups that lack access to standard employment (Graham et al. 2017). On the other hand, gig platforms have become associated with algorithmic surveillance, poorer working conditions that come with on-demand work (that is, employment uncertainty, irregular earnings and unstable working hours), and fewer social rights and voice for workers (Scholz 2017). According to this perspective, many platforms exploit disadvantaged groups by expanding informalized precarious work (Van Doorn 2017; Webster 2016).

In an effort to uphold the benefits and mitigate the risks of gig platforms, discussions about the future of work are on the rise. This chapter builds on insights from sociology and social history to examine how gig workers themselves might strive for more decent work.[1] Since labour market flexibi-

---

[1] 'Decent work' is a concept introduced by the International Labour Organization (ILO) in 1999, later also adopted in the Sustainable Development Goals by the United Nations in 2015, which emphasizes the creation of jobs of acceptable quality (Ghai 2003).

lisation is typically associated with individualisation (Jansen and Akkerman 2014), and platform-mediated work is highly flexible, individual attempts at improving work conditions would be expected on gig platforms. Gig workers are indeed atomised, by competition for short-term jobs and by geographical and social separation from co-workers, which would render them ineffective in taking joint action owing to a lack of solidarity and places to self-organise (Lehdonvirta 2016). In addition, since unionising and collective bargaining are closely tied to the standard employment relationship (Breman and Van der Linden 2014), gig workers are excluded from this collectivism.

Against the odds, empirical observations suggest that various forms of workers' collective action are employed in the platform economy, ranging from street protests against Uber to online community groups by MTurk workers (Graham and Shaw 2017; Johnston and Land-Kazlauskas 2018; Vandaele 2018). Even though collective action by gig workers seems unlikely in theory (Alkhatib et al. 2017), it does occur in practice. This chapter takes that puzzle as its starting point by exploring how collective action in the platform economy differs from more traditional forms of collective action so that expectations on future chances can be derived. I zoom in on a specific type of collective action in the platform economy: worker-owned gig platforms. While most gig platforms are owned by venture capital investors and managed on their behalf, the platform cooperativism movement claims that, together, gig workers can own and govern entirely alternative platforms (Scholz and Schneider 2016). Worker-owned gig platforms are therefore considered an extreme case of collective action, where gig workers control an entirely different arrangement outside of investor-owned platforms that grants them access to working conditions and shared benefits they would otherwise not have.

This chapter makes at least three contributions to the existing literature. First, a brief review of the literature and examples from the field are used to distinguish worker-owned gig platforms from closely related forms of collective action. It thereby tackles conceptual confusion. Second, I place the recent emergence of worker-owned gig platforms in a historical perspective of worker cooperativism. Why do we observe a rediscovery of worker cooperatives now, and why specifically in the platform economy? Third, a comparison between worker-owned gig platforms and traditional worker cooperatives is used to draw out theoretical insights on the opportunities and pitfalls of collective action in the platform economy. Finally, the chapter presents two approaches to collective action in the platform economy and reflects on directions for future research.

MAIN SCIENTIFIC PROBLEM: VARIETIES OF COOPERATIVISM IN THE PLATFORM ECONOMY

In situations where people would be better off collaborating, they must deal with conflicting individual interests that might keep them from doing so. Otherwise the outcome will be suboptimal for both group and individual. For example, food delivery cyclists involved in protests for better working conditions might get banned from the platform and be replaced by more disadvantaged workers for whom accepting such conditions is still a rational decision (Tassinari and Maccarrone 2017). Sociologists describe these situations as collective action problems or social dilemmas (Olson 1971). In order to solve them, people may turn to government regulation or private ownership, but they could also opt for community self-organisation to structure interaction between members of the group (Ostrom 1990). Decent work as a collective good is usually governed by a combination between the three solutions; tripartism between state, employers and labour unions (Berend 2006). This is less evident in the platform economy where non-standard employment and multinational operation restricts the role of states and unions, while most gig workers are also unable to exercise full control over their work as true self-employed.

Since it is unclear how workers' collective action in the platform economy functions exactly, this chapter takes worker-owned gig platforms as a far-reaching example of community self-organisation. We would expect collective action problems to feature heavily in this case, not least because the associated costs of starting or participating in a gig platform that is run by its workers make free-riding behaviour highly attractive. For instance, a member who brings in fewer transaction fees to a worker-owned gig platform may still enjoy better working conditions while the more productive members contribute disproportionally, thus incentivising the latter to prefer investor-owned platforms. To fully understand their collective action problems, an unambiguous definition of worker-owned gig platforms is first required. As there is none, a working definition is derived from literature review and field examples.

The origin of worker-owned gig platforms can be traced to the platform cooperativism movement. Platform cooperatives combine the democratic member-owned-and-governed organisational form of a cooperative (Scholz 2017) with the digital technology and business model of a platform to mediate social and economic interaction (Kenney and Zysman 2016). Various authors have reflected on the possible meanings, designs and preconditions of platform cooperativism (Scholz and Schneider 2016). Although scientific studies are still scarce, the idea of platform cooperativism has been well received in a body of literature that critiques the current state of the platform economy and reverts back to its original associations with sharing and collaborating

(Acquier et al. 2017; Gruszka 2017; Sarina and Riley 2018; Schor 2014). Frenken (2017) suggests that platform cooperatives could leverage knowledge and (legal) recognition from the two centuries long history of cooperativism, for instance, by merging this history with the new cooperative developments in the platform economy.

Two theoretical studies discuss worker-owned gig platforms in more detail. Belloc (2018) applies mathematical modelling to compare the per-capita earnings of a worker-owned gig platform to an investor-owned platform under various hypothetical conditions. He concludes that investor-owned platforms remain more attractive to workers up to a particular threshold of overhead costs they have to pay to the platform owner on each transaction. Nonetheless, per-capita earnings may not be the only factor that workers take into consideration when choosing between an investor-owned and a worker-owned gig platform (if they are able to make a choice between these alternatives in the first place). Arets and Frenken (2019) predict that the cooperative model will be more feasible for some sectors of platform-mediated service provision than for others. In particular, social care and cleaning services, as these are local markets, require simple platform functionalities, feature recurrent interactions with customers, provide main incomes and there are existing cooperatives in these sectors (at least in the Netherlands). Both studies have in common that worker-owned gig platforms are discussed as alternatives to investor-owned gig platforms.

In contrast, there are also instances in which worker cooperatives collaborate with investor-owned platforms. Drahokoupil and Piasna (2019) carried out a case study on the arrangement between food delivery platform Deliveroo and worker cooperative Smart in Belgium between 2016 and 2018. Before this arrangement was ultimately abandoned by Deliveroo, workers could gain employment status and associated social rights through the cooperative while effectively still working for the investor-owned platform. Although the researchers state that Smart could be criticised for normalising precarious work, most riders in their survey preferred working through Smart and experienced a loss of autonomy when they returned to the status quo of self-employment after Deliveroo had abandoned the arrangement. The benefits of these arrangements have been noted by legal scholars as well (Van Slooten and Holscher 2019), for instance, by collective pooling of equipment, insurance and training, though they fail to classify as worker-owned gig platforms because the platform itself is not worker owned and governed.[2]

---

[2] Smart also provides a job board platform itself, but this was not used by the riders.

Other distinctions could be made as well. Como et al. (2016) distinguish digital innovation for the internal organisation of cooperatives (for example, to facilitate members' participation in decision-making) from actual platform cooperatives that use digital innovation for their business model (for studies on the former, see Jackson and Kuehn 2016; Mannan 2018). In addition, we could distinguish worker-owned gig platforms as being software-based from worker cooperatives that produce digital hardware (Schneider 2018). Building on Como et al. (2016, p. 9), worker-owned gig platforms can be established through three processes that demand different levels of effort from workers' collective action:

1. Creation – a new worker-owned gig platform is built and organised from scratch (for example, Pwiic, which was founded in 2017).
2. Conversion – an existing platform is mutualised by its workers (no known examples).[3]
3. Coding – an existing worker cooperative adopts a platform (for example, Auckland Co-Op Taxis, which had existed since 1947 before adopting an application (app) similar to Uber's in 2013).

Both platforms and cooperatives come in many different shapes. Following Eum (2017, p. 31), it is possible to distinguish four meta-types of cooperatives based on who their members are; that is, user, producer, worker and multi-stakeholder. Users, often consumers, set up cooperatives to benefit from economies of scale in the collective procurement of goods and services that they otherwise find difficult to obtain. User cooperatives are not directly related to jobs or production. Various types of producer-members (for example, farmers, artisans and entrepreneurs) set up cooperatives to collectively acquire shared services directly related to production, such as joint storage, marketing and distributing. Workers use a cooperative for access to employment, control over work conditions and sharing in profit. Multi-stakeholder or hybrid cooperatives involve a mix of member classes to strike a balance between different interests. However, theoretically multi-stakeholder cooperatives should face more stringent collective action problems owing to their size and heterogeneous interests (Olson 1971).

It should be noted, however, that distinctions between consumer, producer and worker can become fuzzy in the platform economy. For instance, Facebook users simultaneously consume the platform's social networking services while producing content on which advertising revenue is gained (that is, prosump-

---

[3]   Mutualisation here means (re)organising an enterprise so that it is owned and democratically governed by its workers. Inversely, demutualisation implies a shift to investor ownership.

Table 9.1   *Examples of platform cooperatives used in classification*

Name	Year of launch*	Member classes	Country/ region	Business model
Modo	2002	Individual and business car users	Vancouver, BC	Sharing a fleet of cars
Fairmondo	2013	Sellers; consumers; platform staff	Germany; the United Kingdom	Sale of ethical goods and artistic products
Stocksy United	2013	Artists; platform staff; advisors	Canada; worldwide	Sale of stock photography and cinematography
Loconomics	2016	Freelance service professionals; platform staff; worker cooperatives; non-profit organisations; licensing partners that wish to use the platform technology	San Francisco, CA	Service provision (from web design to dog-walking)
Resonate	2017	Artists; music labels; listeners; platform staff	Germany; worldwide	Music streaming service
RidyGo	2017	Users; platform staff	Cote d'Azur, France	Carpooling, especially for unemployed jobseekers
CoopCycle	2018	Cooperatives of deliverers (majority share); restaurant/ shop-owners; platform staff; public partners; private partners	France; Spain; Belgium; Germany	Bicycle food delivery/ courier services
Fairbnb	2019	Hosts; guests; local business owners; neighbours	Italy; Spain; the Netherlands	Short-term vacation rentals

Note: * Founding dates (in year of launch column) of these platform cooperatives are typically earlier than the launch date. For example, Modo operated from 1997 using trip-logbooks before adopting an online platform.

tion). Moreover, the services provided on gig platforms are not executed by its employed technical and managerial staff, but by workers who are commonly self-employed (Schneider 2018). This is why most existing platform cooperatives involve multiple member classes (see Table 9.1) and previous literature on platform cooperativism has made little effort to distinguish between different types of cooperatives (for an exception, see Borkin 2019).

I argue, instead, that it is not useful to lump all forms of platform cooperativism together. Different types of cooperatives have different functions underlying their existence and, without classification, it becomes impossible to make meaningful comparisons (Zamagni 2012). Following Borkin (2019), I argue that platforms can be more or less labour intensive. For instance, a platform that facilitates short-term vacation rentals is not as labour intensive as a platform for home cleaning. So, even when most platform cooperatives are open to multiple member classes, particular platform categories are a better match with worker cooperativism than others. Therefore, only platform cooperatives that facilitate a labour exchange between service providers and customers as their core business may be qualified as worker-owned gig platforms (Loconomics and CoopCycle in Table 9.1). Although some scholars have questioned the ability of platform cooperatives to move beyond the trade in ethical goods and artistic products (Belloc 2018), the examples in Table 9.1 illustrate a larger diversity and opportunities for worker-owned gig platforms specifically. Note that, following the International Labour Organization's (2016) definition for non-standard employment, the term 'worker' here is based more on their dependency on the platform than their status of employee versus self-employed.

Based on the above literature review and classification of field examples, I write of a worker-owned gig platform when: (1) the fulfilment of worker-specific needs is a central aim of the cooperative, (2) the platform is primarily worker owned and governed, and (3) the platform intermediates paid labour exchange. The first criterion relates to the purpose of worker cooperatives, both historically and legally (Zamagni 2012). The second criterion distinguishes worker-owned gig platforms from arrangements between worker cooperatives and investor-owned platforms (Van Slooten and Holscher 2019). The third criterion distinguishes worker-owned gig platforms from other types of platform cooperatives that might even include a worker member class, but do not use platform technologies to intermediate service provision by workers to clients (Borkin 2019).

CAUSES OF THE REDISCOVERY OF WORKER COOPERATIVES

Interest in worker cooperatives increases with the recent popularity of social entrepreneurship and flattened organisations in management discourse, the 2008 economic crisis and the United Nations declaring 2012 the International Year of Cooperatives (Cheney et al. 2014). The history of worker cooperativism, however, dates back to at least the nineteenth century. This section places worker-owned gig platforms in a historical perspective of worker cooperativism. Why do we observe a rediscovery of worker cooperatives now, and why specifically in the platform economy?

Table 9.2   *Historical phases of worker cooperativism*

	Phase 1: Early industrial period	Phase 2: Later industrial period	Phase 3: Post-industrial period
Dominant political-economic system	Pre-Fordism/liberal capitalism	Fordism/organised capitalism	Post-Fordism/disorganised capitalism
New goal of worker cooperatives	Uplift the working class to a decent standard of life	Counteract deskilling and increase workplace democratic participation	Combine individual autonomy with the security of a workers' collective
Emblematic case	Hebden Bridge Fustian Manufacturing Society	Mondragón Cooperative Corporation	Smart

Although a detailed historical treatise on worker cooperatives is beyond the scope of this chapter,[4] I distinguish three phases of worker cooperativism over the past two centuries (see Table 9.2). By examining the changing macro-level circumstances that affect individual considerations to start or participate in workers' collective action, it becomes apparent for each phase how the (re) discovery of worker cooperatives is associated with needs that are unsatisfied by the private and public sector (Moulaert and Ailenei 2005).

The first phase of worker cooperativism manifests itself especially in the second half of the nineteenth century as a reaction against poverty and exploitation during the Industrial Revolution (Rothschild 2009), but also amid the power vacuum between the dawn of guilds and the rise of states as economic actors (Moulaert and Ailenei 2005). That is, there was both demand and opportunity for collective action. It is first and foremost an experimental phase in which worker cooperatives lacked legal recognition. Manufacturing jobs attracted people to cities, but worker protections and social rights were still very limited under liberal capitalism (Goldkind and McNutt 2019). Inspired by the cooperative communities of utopian socialists[5] and the famous Rochdale

---

[4]   Bear in mind that this is an analytical typology, so phases may overlap and vary locally in practice. Moreover, the typology has a European focus and does not include proto-cooperatives. Worker cooperatives in socialist states are also beyond the scope of this chapter.

[5]   These cooperative communities are different from modern cooperatives. They typically had ownership and most governance ascribed to a patron-founder (for example, Robert Owen in the example of New Harmony), and aimed to establish settlements that combine cooperative work and consumption.

Principles in 1844,[6] the characteristics of worker cooperativism started to crystallise.

An example here is Hebden Bridge Fustian Manufacturing Society in the United Kingdom, which was just as pioneering to worker cooperativism as the Rochdale Pioneers were to consumer cooperativism (Bibby 2015). It produced heavy cloth and clothing in its own Nutclough Mill, employed 300 women and men at its peak, and ran profitably from its start in 1871 until it was taken over by the Co-operative Wholesale Society in 1918.[7] In the words of its founder, the ultimate purpose of the Fustian Manufacturing Society was 'redemption of the working people' (Bibby 2015, p. 191). This included better wages and a share of the surplus value, co-ownership and a vote in general meetings through shareholding, safer workplace conditions, and education opportunities for workers. Similar to most worker cooperatives in the first phase, however, workers' collective ownership and democratic self-management were not yet applied.

The second phase of worker cooperativism, mostly after World War II, takes place in a Fordist economic context of deskilling and scientific management control, while the Great Depression and pressure from labour movements ushered in a state-regulated capitalism (Lash and Urry 1987). It is a phase during which worker cooperatives formalised and gained legal recognition, but also during which many of their social functions were taken over by the nation state (Moulaert and Ailenei 2005). The standard employment relationship and a more extensive social welfare system were established in many countries. Standard employment refers to a full-time job with one employer over a long time span, guaranteeing regular hours and income for wage earners, but also access to social rights and classical workers' collective action based on unionising and collective bargaining (Breman and Van der Linden 2014). Standardisation in Fordist production processes required less skilled labour, while being more capital intensive. These circumstances resulted in some researchers describing the twentieth century as 'no place for [worker] cooperatives' (Zamagni 2012, p. 31). The opportunities for this specific form of collective action decreased because much of the demand was already fulfilled by states and labour unions. Nonetheless, it was also during this period that

---

[6]   The seven principles of the Rochdale pioneers have since become a guideline to all cooperatives, as agreed by the International Co-operative Alliance (n.d.): (1) voluntary and open membership; (2) democratic member control; (3) member economic participation; (4) autonomy and independence; (5) education, training and information; (6) cooperation among cooperatives; and (7) concern for community.

[7]   A wider dispute is noted during this time between worker and consumer cooperativism as separate strategies to improve living conditions for the working class (Webb 1904).

state support allowed new worker cooperatives to start and some of the largest worker cooperatives of all time started up (Rothschild 2009).

Other considerations for starting or joining a worker cooperative became important during the second phase. The Mondragón Cooperative Corporation from the Basque province of Spain, founded in 1956, is a well-known example (Whyte and Whyte 1991). As a federation of worker cooperatives, it still exists and includes over 100 organisations in manufacturing, finance, retail and knowledge. The prominent role of education in Mondragón contrasts with the deskilling of workers common to Fordist production. In addition, worker cooperatives of the second phase provided a vehicle for political expression (Baskaran 2015). Reflecting the wider cultural shifts and social movements of the 1960s, empowerment through workers' collective ownership and democratic self-management became increasingly central to worker cooperativism (Gupta 2014). This aimed to replace the increasing bureaucracy of that time with workplace democracy (Rothschild and Russell 1986).

The third phase of worker cooperativism, starting near the end of the twentieth century, can be seen as a response to the retreating welfare state and the transition towards a post-industrial economy (Moulaert and Ailenei 2005). It is a phase during which worker cooperatives re-emerge as a form of collective action (Baskaran 2015; Cheney et al. 2014). Especially in times of economic crises, such as the 2008 financial crisis, the demand for worker cooperatives arises from needs that are unfulfilled by a retreating welfare state and declining unionism (Moulaert and Ailenei 2005). Globalisation and the rise of service sectors triggered a fall of the standard employment relationship in favour of non-standard employment (Stanford 2017). However, non-standard employment also tends to exclude workers from labour protections and social rights precisely because the classic model of collective action is so strongly connected to standard employment (Goldkind and McNutt 2019). It should be noted, however, that what is termed 'standard' and 'classic' here is now often reinterpreted by social historians as an exceptional period in the twentieth century that especially served male breadwinners in Western countries (Breman and Van der Linden 2014). New considerations for starting or participating in, a worker cooperative involve the desire to unite a secure livelihood with the flexibility and individual autonomy of non-standard employment (Schneider 2018). This is reflected in worker-owned gig platforms, but not exclusively so. For instance, social cooperatives help unemployed people to reintegrate into the labour market through entrepreneurial activities (Borzaga 1996).

An example of this third phase is the previously discussed worker cooperative, Smart. Founded in Belgium to provide administrative services to artists in 1998, but converted to a formal cooperative structure in 2016, Smart now has over 80 000 members in eight European countries and is open to all types of non-standard work (Drahokoupil and Piasna 2019).

> It invoices contractors on behalf of its [...] members and, after deducting a fee and the respective taxes and contributions, pays salary to its members. It thus offers administrative, accounting, and financial services to its members, who otherwise face a complex regulatory environment, and allows them to access social security through the employment contract. (Drahokoupil and Piasna 2019, p. 7)

Whether these new initiatives still qualify as worker cooperatives in the narrow sense is a point of discussion. Smart is sometimes also described as a business and employment cooperative, which provides freelancers the position of salaried employee including social protections but also the independence of self-employment.

CONSEQUENCES FOR COLLECTIVE ACTION BY GIG WORKERS

In order to understand the implications for collective action problems in the platform economy, I compare traditional worker cooperatives as they have existed historically with the recently emerging worker-owned gig platforms (see Table 9.3). A comparison is made along three dimensions of institutions for collective action: members, resource, and institution (De Moor 2015).

Table 9.3   *Comparison between traditional worker cooperatives and worker-owned gig platforms to derive opportunities (+) and pitfalls (−) for collective action in the platform economy*

	Traditional worker cooperatives	Worker-owned gig platforms	Implications for collective action in the platform economy
Members	Mostly small group size, socially homogenous and geographically local	Potentially large group size, socially heterogeneous and geographically dispersed	+ Scaling − Fragmentation
Resource	Standard employment, relatively more often manufacturing jobs, team production	Non-standard employment, relatively more often service jobs, solo production	+ Capital conundrum − Competition
Institution	Coordination by elected managers	Coordination by elected algorithms	+ Self-governance − Digital divides

First, if we look at the members of these collective actions, the workers, a few implications can be derived. Apart from some exceptions, members of traditional worker cooperatives are often (by design) small in number, socially homogenous and part of the same local communities (Benham and Keefer 1991; Gupta, 2014). Even though research shows this is not so different for most capitalist firms (Pérotin 2013), it has a different meaning for worker cooperatives where trust and reciprocity are indispensable without the presence of clear hierarchy. Worker-owned gig platforms have potential for larger size, social heterogeneity and geographical dispersion. This is mainly because Internet technologies reduce the cost of communicating over long distances and with many people (Lupia and Sin 2003). Some types of platform-mediated work can even be performed remotely from behind a computer anywhere in the world (De Stefano 2015) and the diversity of services provided on some gig platforms increases workers' social heterogeneity as well (for example, education, experience and skills).

On the one hand, this implies an upscaling from collective action by very exclusive worker communities to inclusive collective action for decent work as a common good. With a larger number and more diverse group of workers, it is relatively easy to achieve economies of scale that can cover collective costs (Oliver and Marwell 1988). This holds especially true for worker-owned gig platforms, as indirect network externalities make a platform more attractive to customers as the number of service providers increases (Frenken et al. 2017). On the other hand, there may be higher costs associated with the fragmentation that comes with a larger group size, social heterogeneity and geographical dispersion. This is because social control and solidarity are harder to establish when individual behaviour is not visibly monitorable or there are diverging interests among workers, such as between those that are dependent on the platform for their main income and those that only supplement their income (Frenken 2017). Building a shared identity and social control is not ruled out entirely, but the costs of exercising it become higher, especially for crowdwork.

Second, labour as a central resource to both forms of collective action differs considerably. Traditionally, worker cooperatives provide full-time jobs to their members by means of a standard employment contract (Eum 2017). This also makes worker-members eligible for various social rights in welfare states. Labour tasks are typically performed together with co-workers and shared means of production, that is, team production (Dow 2003). Worker-owned gig platforms almost exclusively provide service jobs to their members and employment can be non-standard in multiple ways (Eum 2017), such as part-time with a contract or self-employment. Labour tasks of gig workers are typically performed alone and with personally owned means of production, that is, solo production (De Stefano 2015).

On the plus side, these characteristics enable collective action in the platform economy to overcome the capital conundrum more easily than was historically possible. One reason for this is that service jobs are less capital intensive, thus requiring fewer large investments (Zamagni 2012). Also, despite of research showing worker cooperatives to be no less capital intensive than capitalist firms (Pérotin 2013), the people who undertake collective action in the platform economy may be less wealthy than those who start an investor-owned platform. In addition, the flexibility of non-standard employment and personally owned means of production lower the cost of creating jobs (Breman and Van der Linden 2014); though it should be noted that there is a trade-off between lowering these costs and the benefits that workers pool together, which makes collective action attractive in the first place. Looking at pitfalls, the shift towards solo production, and in some cases self-employment, allows for productivity differences and competition between workers (Schor et al. 2020). In turn, this can create social dilemmas where individuals benefit from selfish behaviour while others still contribute. Such a perverse incentive makes it less likely that collective action by gig workers will provide benefits such as knowledge or risk-sharing, unless measures are taken to obligate particular behaviours (for example, monthly fees and boycott of working for investor-owned platforms). Competition between members also raises the question of whether worker-owned gig platforms qualify as new types of worker cooperatives at all. If members are self-employed and mainly use the cooperative for shared services, such as work-finding, the organisation starts to resemble a producer cooperative (Eum 2017). When self-employed gig workers organise to gain access to shared services and make rules to discourage free-riding, tensions with competition law and, in particular, horizontal anti-trust legislation arise (Arets and Frenken 2019). There are exemptions for some producer cooperative types (Grillo 2013), such as in agriculture, but currently it is unclear whether these will hold too for worker-owned gig platforms.

Finally, some differences can be observed when comparing the institutional rules and decision-making between traditional worker cooperatives and worker-owned gig platforms. As democratic member organisations, traditional worker cooperatives usually have a delegated management that is elected by worker-members, with one vote each, to coordinate the organisation (Vieta et al. 2016). Fundamental rules about governance are also laid out in by-laws, a type of constitution for the cooperative, although, at the same time, worker-members could be entitled to vote on major decisions in a general assembly, participate in committees and/or run for management themselves. Worker-owned gig platforms may still have all of this, but also integrate by-laws and other decisions into their algorithms and platform design (Martin et al. 2017). They employ digital tools for governing the organisation on a daily basis as well, ranging from simple message boards to smart contracts that could replace the need for delegated management altogether (Mannan 2018).

Implications for collective action in the platform economy include new opportunities for digital self-governance, but also digital divides between information technology technicians and ordinary people without transparency or control over important algorithms (Schradie 2018). Involvement of all members, not just leaders, in decision-making has always been a challenge for collective action. Classic sociologists such as Max Weber and Robert Michels have theorised that even democratic organisations tend towards hierarchies and power concentration (Rothschild and Whitt 1989). Information technologies potentially enable more direct democratic participation, often described as digital democracy, by making participation in decision-making on larger scales more efficient (Dahlberg 2011). However, there is also the danger that new hierarchies and power concentration will emerge based on technology shrewdness (Eum 2017). Crowdwork platform Daemo, for instance, aimed but ultimately failed to become democratically self-governed because its technical staff from Stanford University held most control (Social Contract CR 2017).

SOLUTIONS TO THE PLATFORM ECONOMY'S COLLECTIVE ACTION PUZZLE

There is a broad range of strategies for gig workers to undertake collective action (Johnston and Land-Kazlauskas 2018), going far beyond the worker-owned gig platforms that are central this chapter. Some examples include street protests, new forms of unionising and online worker forums. However, based on the implications derived in the previous section it becomes possible to predict that all gig workers' strategies for collective action will fall into either of two approaches (see Figure 9.1). Making this distinction also provides some practical insight into the future chances of solving collective action problems around decent work in the platform economy.

The soft approach starts with groups of gig workers who are willingly self-employed and/or more geographically and socially fragmented (Lenaerts et al. 2018). Limitations posed by competition law and international legal differences, but also the undesirability for many freelancers of social control infringing on their own autonomy will incentivise these workers to set only a few rules to structure their collective action; that is, a relatively lower level of institutionalisation. However, the social dilemma of conflicting individual and group interests then becomes more pressing (Olson 1971). Free-riding on the effort of other gig workers in collective action is more attractive when there are few rules to be monitored and enforced. For example, street protests by Foodora delivery cyclists did not result in the company meeting their core demands and, while protesters' accounts were deactivated, a large pool of labour easily

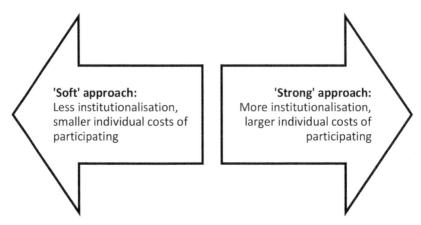

Figure 9.1   Two main approaches to gig workers' collective action in the platform economy

replaced them with new workers (Tassinari and Maccarrone 2017). To evade collective action problems, it is expected that the soft approach will keep individual costs of participating low and have little obligations for members. The utility of this approach should not be underestimated though. Studies on online community groups of gig workers show that, even without sharing much collectively, just information and social connections, can increase gig workers' protection (Irani and Silberman 2013; Salehi et al. 2015; Wood et al. 2018); for instance, by building alternative reputation systems on good and bad clients or through creating social capital with other workers.

In contrast, the strong approach starts with groups of gig workers who already have some type of employment contract or who want a contract, and are less fragmented owing to geographical or social distance between them (Lenaerts et al. 2018). These workers face fewer legal restrictions to organising together, setting up rules and making agreements. They may also experience more of a common identity and thus be willing to accept social control by the collective (Lehdonvirta 2016). Levels of institutionalisation are then relatively higher (Ostrom 1990). Since opportunistic behaviour by individual gig workers is less likely with rules that are monitored and enforced, workers are incentivised to raise the stakes of collective action. Therefore, it is expected that the second approach addresses collective action problems head-on instead of merely evading them. Studies on traditional and new labour unions that engage with gig workers show results in minimum pay, discounted insurance, safety guidelines and dispute settlement (Johnston and Land-Kazlauskas 2018; Minter 2017). Worker-owned gig platforms are also likely to follow this second approach, but might opt for the first if they minimise what is organised collectively for work-finding, information exchange and social connection through their platform, for which they pay only maintenance costs.

DIRECTIONS FOR FUTURE RESEARCH

This chapter presented the puzzle of gig workers' collective action, which is a phenomenon that should theoretically be unlikely to occur but is still observed in practice (Alkhatib et al. 2017; Johnston and Land-Kazlauskas 2018). A comparison of worker-owned gig platforms with traditional worker cooperatives was used to derive insights on how collective action problems may be solved in the platform economy through community self-organisation. This section presents an agenda for future research.

First, it should be emphasised that worker-owned gig platforms are not isolated phenomena. Community self-organising is just one of the three main solutions to collective action problems. Future studies should therefore assess how community self-organisation by gig workers is affected by the state and market that surrounds it. Collective goods are often produced in a context of institutional diversity (Ostrom 1990), which is no different for decent work in the platform economy. Variation in laws and policies between countries could influence both individual gig workers' demand for collective action as well as the opportunities and pitfalls for collective action in the platform economy. The same holds for variation by industry or type of work. Sociologists are well equipped to examine these issues through cross-country comparisons and by comparing different groups of gig workers.

Second, while focusing on worker-owned gig platforms as an extreme case of collective action in the platform economy helped to bring out its central problems, it also probably coloured the implications that are drawn about gig workers' collective action. Further research should therefore study different instances in a similar manner, for example, by comparing independent worker unions directed at gig workers with traditional labour unions (Lenaerts et al. 2018) or 'crowd guilds' with historical craft guilds (Whiting et al. 2017). Another approach would be to look at how a particular occupational group has organised itself over time and does so currently in the platform economy. Social history is rich with collective action by workers, and platform economy researchers underutilise the potential of learning from the past by analysing both similarities and differences.

Finally, without downplaying the importance of theoretical work, there is still a lack of empirical research on collective action in the platform economy. Some pioneering explorative studies show that demand for collective action is probably clustered among particular subsets of gig workers (Fellmoser 2018; Newlands et al. 2018) and that a variety of collective actions is currently undertaken (Graham and Shaw 2017; Johnston and Land-Kazlauskas 2018; Vandaele 2018). However, future studies should direct attention to the viability of gig workers' collective action by analysing questions of scaling, fragmentation, capital conundrum, competition, self-governance and digital divides. Expectations can then be tested from the soft and strong approaches, as described in the previous section, to gain an understanding of how gig workers overcome (or fail to overcome) collective action problems.

ACKNOWLEDGEMENTS

This study is part of the research program Sustainable Cooperation – Roadmaps to Resilient Societies (SCOOP). The author is grateful to the Netherlands Organization for Scientific Research (NWO) and the Dutch Ministry of Education, Culture and Science (OCW) for generously funding this research in the context of its 2017 Gravitation Program (grant number 024.003.025).

REFERENCES

Acquier, A., T. Daudigeos and J. Pinkse (2017), 'Promises and paradoxes of the sharing economy: an organizing framework', *Technological Forecasting and Social Change*, **125** (December), 1–10.

Alkhatib, A., M.S. Bernstein and M. Levi (2017), 'Examining crowd work and gig work through the historical lens of piecework', in *Proceedings of the 2017 CHI Conference on Human Factors in Computing Systems*, New York: Association for Computing Machinery, pp. 4599–616.

Arets, M.G.C. and K. Frenken (2019), 'Zijn platformcoöperaties levensvatbaar?' ('Are platform cooperatives viable?') *TPEdigitaal*, **13** (2), 1–14.

Baskaran, P. (2015), 'Introduction to worker cooperatives and their role in the changing economy', *Journal of Affordable Housing and Community Development Law*, **24** (2), 355–81.

Belloc, F. (2018), 'Why isn't Uber a worker cooperative? On the (im)possibility of self-management in the platform economy', Società Italiana degli Economisti, accessed 31 December 2018 at https://siecon3-607788.c.cdn77.org/sites/siecon.org/files/media_wysiwyg/belloc-81.pdf.

Benham, L. and P. Keefer (1991), 'Voting in firms: the role of agenda control, size and voter homogeneity', *Economic Inquiry*, **29** (4), 706–19.

Berend, I.T. (2006), *An Economic History of Twentieth-Century Europe: Economic Regimes from Laissez-Faire to Globalization*, Cambridge: Cambridge University Press.

Bibby, A. (2015), *All Our Own Work: The Co-operative Pioneers of Hebden Bridge and Their Mill*, London: Merlin Press.

Borkin, S. (2019), *Platform Co-operatives – Solving the Capital Conundrum*, London and Manchester: Nesta, Co-operatives UK.

Borzaga, C. (1996), 'Social cooperatives and work integration in Italy', *Annals of Public and Cooperative Economics*, **67** (2), 209–34.

Breman, J. and M. Van der Linden (2014), 'Informalizing the economy: the return of the social question at a global level', *Development and Change*, **45** (5), 920–40.

Cheney, G., I. Santa Cruz, A.M. Peredo and E. Nazareno (2014), 'Worker cooperatives as an organizational alternative: challenges, achievements and promise in business governance and ownership', *Organization*, **21** (5), 591–603.

Como, E., A. Mathis, M. Tognetti and A. Rapisardi (2016), 'Cooperative platforms in a European landscape: an exploratory study', paper presented at the ISIRC Conference, Glasgow – Cooperatives Europe, September, accessed 27 December 2018 at https://coopseurope.coop/resources/news/collaborative-economy -opportunity-cooperatives-new-study-released.

Dahlberg, L. (2011), 'Re-constructing digital democracy: an outline of four "positions"', *New Media and Society*, **13** (6), 855–72.

De Moor, T. (2015), *The Dilemma of the Commoners: Understanding the Use of Common-Pool Resources in Long-Term Perspective*, New York: Cambridge University Press.

De Stefano, V. (2015), 'The rise of the just-in-time workforce: on-demand work, crowdwork, and labor protection in the gig-economy', *Comparative Labor Law and Policy Journal*, **37** (3), 471–503.

Dow, G.K. (2003), *Governing the Firm: Workers' Control in Theory and Practice*, Cambridge: Cambridge University Press.

Drahokoupil, J. and A. Piasna (2019), 'Work in the platform economy: Deliveroo riders in Belgium and the smart arrangement', ETUI Working Paper No. 2019.01, accessed 7 September 2020 at https://ssrn.com/abstract=3316133.

Eum, H. (2017), *Cooperatives and Employment, Second Global Report 2017*, Brussels: International Cooperative Alliance.

Fellmoser, M. (2018), 'Out of place out of sight? A quantitative study on social connectedness in the platform economy and its effect on the willingness to participate in collective action', BA thesis, University of Twente, accessed 7 September 2020 at http://essay.utwente.nl/75858/1/Fellmoser%20BA%20BMS.pdf.

Frenken, K. (2017), 'Political economies and environmental futures for the sharing economy', *Philosophical Transactions of the Royal Society A*, **375** (2095), 1–15.

Frenken, K., A.H.M. Van Waes, M.M. Smink and R. Van Est (2017), *A Fair Share – Safeguarding Public Interests in the Sharing and Gig Economy*, The Hague: Rathenau Institute.

Ghai, D. (2003), 'Decent work: concept and indicators', *International Labour Review*, **142** (2), 113–45

Goldkind, L. and J.G. McNutt (2019), 'Vampires in the technological mist: the sharing economy, employment and the quest for economic justice and fairness in a digital future', *Ethics and Social Welfare*, **13** (1), 51–63.

Graham, M., I. Hjorth and V. Lehdonvirta (2017), 'Digital labour and development: impacts of global digital labour platforms and the gig economy on worker livelihoods', *Transfer: European Review of Labour and Research*, **23** (2), 135–62.

Graham, M. and J. Shaw (eds) (2017), *Towards a Fairer Gig Economy*, London: Meatspace Press.

Grillo, M. (2013), 'Competition rules and the cooperative firm', *Journal of Entrepreneurial and Organizational Diversity*, **2** (1), 36–53.

Gruszka, K. (2017), 'Framing the collaborative economy – voices of contestation', *Environmental Innovation and Societal Transitions*, **23** (June), 92–104.

Gupta, C. (2014), 'The co-operative model as a "living experiment in democracy"', *Journal of Co-operative Organization and Management*, **2** (2), 98–107.

International Co-operative Alliance (n.d.), 'Cooperative identity, values & principles', accessed 7 September 2020 at https://www.ica.coop/en/cooperatives/cooperative-identity.

International Labour Organization (2016), *Non-Standard Employment Around the World: Understanding Challenges, Shaping Prospects*, Geneva: International Labour Office.

Irani, L.C. and M. Silberman (2013), 'Turkopticon: interrupting worker invisibility in Amazon Mechanical Turk', in *Proceedings of the SIGCHI Conference on Human Factors in Computing Systems*, New York: Association for Computing Machinery, pp. 611–20.

Jackson, S.K. and K.M. Kuehn (2016), 'Open source, social activism and "necessary trade-offs" in the digital enclosure: a case study of platform co-operative, Loomio. Org', *tripleC: Communication, Capitalism and Critique. Open Access Journal for a Global Sustainable Information Society*, **14** (2), 413–27.

Jansen, G. and A. Akkerman (2014), 'The collapse of collective action? Employment flexibility, union membership and strikes in European companies', in M. Hauptmeier and M. Vidal (eds), *Comparative Political Economy of Work*, London: Palgrave Macmillan, pp. 186–207.

Johnston, H. and C. Land-Kazlauskas (2018), *Organizing On-Demand: Representation, Voice, And Collective Bargaining in the Gig Economy*, Geneva: International Labour Office.

Kenney, M. and J. Zysman (2016), 'The rise of the platform economy', *Issues in Science and Technology*, **32** (3), 61–9.

Lash, S. and Urry, J. (1987), *The End of Organized Capitalism*, Madison, WI: University of Wisconsin Press.

Lehdonvirta, V. (2016), 'Algorithms that divide and unite: delocalisation, identity and collective action in "microwork"', in J. Flecker (ed.), *Space, Place and Global Digital Work*, London: Palgrave Macmillan, pp. 53–80.

Lenaerts, K., Z. Kilhoffer and M. Akgüç (2018), 'Traditional and new forms of organisation and representation in the platform economy', *Work Organisation, Labour and Globalisation*, **12** (2), 60–78.

Lupia, A. and G. Sin (2003),' Which public goods are endangered? How evolving communication technologies affect the logic of collective action', *Public Choice*, **117** (3–4), 315–31.

Mannan, M. (2018), 'Fostering worker cooperatives with blockchain technology: lessons from the Colony Project', *Erasmus Law Review*, **11** (3), 190–203.

Martin, C.J. (2016), 'The sharing economy: a pathway to sustainability or a nightmarish form of neoliberal capitalism?', *Ecological Economics*, **121** (January), 149–59.

Martin, C.J., P. Upham and R. Klapper (2017), 'Democratising platform governance in the sharing economy: an analytical framework and initial empirical insights', *Journal of Cleaner Production*, **166** (November), 1395–406.

Minter, K. (2017), 'Negotiating labour standards in the gig economy: Airtasker and Unions New South Wales', *Economic and Labour Relations Review*, **28** (3), 438–54.

Moulaert, F. and O. Ailenei (2005), 'Social economy, third sector and solidarity relations: a conceptual synthesis from history to present', *Urban Studies*, **42** (11), 2037–53.

Newlands, G., C. Lutz and C. Fieseler (2018), 'Collective action and provider classification in the sharing economy', *New Technology, Work and Employment*, **33** (3), 250–67.

Oliver, P.E. and G. Marwell (1988), 'The paradox of group size in collective action: a theory of the critical mass, II', *American Sociological Review*, **53** (1), 1–8.

Olson, M. (1971), *The Logic of Collective Action: Public Goods and the Theory of Groups*, revd edn, Cambridge, MA: Harvard University Press.

Ostrom, E. (1990), *Governing the Commons: the Evolution of Institutions for Collective Action*, New York: Cambridge University Press.

Pérotin, V. (2013), 'Worker cooperatives: good, sustainable jobs in the community', *Journal of Entrepreneurial and Organizational Diversity*, **2** (2), 34–47.

Rothschild, J. (2009), 'Workers' cooperatives and social enterprise: a forgotten route to social equity and democracy', *American Behavioral Scientist*, **52** (7), 1023–41.

Rothschild, J. and R. Russell (1986), 'Alternatives to bureaucracy: democratic participation in the economy', *Annual Review of Sociology*, **12** (1), 307–28.

Rothschild, J. and J.A. Whitt (1989), *The Cooperative Workplace: Potentials and Dilemmas of Organisational Democracy and Participation*, Cambridge: Cambridge University Press.

Salehi, N., L.C. Irani, M.S. Bernstein, A. Alkhatib, E. Ogbe and K. Milland (2015), 'We are dynamo: overcoming stalling and friction in collective action for crowd workers', in *Proceedings of the 33rd Annual ACM Conference on Human Factors in Computing Systems*, New York: Association for Computing Machinery, pp. 1621–30.

Sarina, T. and J. Riley (2018), 'Re-crafting the enterprise for the gig-economy', *New Zealand Journal of Employment Relations*, **43** (2), 27–35.

Schneider, N. (2018), 'An internet of ownership: democratic design for the online economy', *Sociological Review*, **66** (2), 320–40.

Scholz, T. (2017), *Uberworked and Underpaid: How Workers are Disrupting the Digital Economy*, Cambridge: Polity Press.

Scholz, T. and N. Schneider (eds) (2016), *Ours to Hack and to Own: The Rise of Platform Cooperativism, a New Vision for the Future of Work and a Fairer Internet*, New York: OR Books.

Schor, J.B. (2014), 'Debating the sharing economy', *Great Transition Initiative*, accessed 27 December 2018 at http://www.greattransition.org/images/GTI _publications/Schor_Debating_the_Sharing_Economy.pdf.

Schor, J.B., M. Cansoy, W. Charles, L. Ladegaard and R. Wengronowitz (2020), 'Dependence and precarity in the platform economy', *Theory and Society*, **49** (5), 833–61.

Schradie, J. (2018), 'The digital activism gap: how class and costs shape online collective action', *Social Problems*, **65** (1), 51–74.

Social Contract CR (2017), 'Daemo's crowd has had enough', *Medium*, accessed 7 September 2020 at https://medium.com/@SocialContractCR/daemos-crowd-has -had-enough-5b62c1fceb30.

Stanford, J. (2017), 'The resurgence of gig work: historical and theoretical perspectives', *Economic and Labour Relations Review*, **28** (3), 382–401.

Tassinari, A. and V. Maccarrone (2017), 'The mobilisation of gig economy couriers', *Transfer: European Review of Labour and Research*, **23** (3), 353–7.

Vandaele, K. (2018), 'Will trade unions survive in the platform economy? Emerging patterns of platform workers' collective voice and representation in Europe', ETUI Working Paper No. 2018.05, doi:/10.2139/ssrn.3198546.

Van Doorn, N. (2017), 'Platform labor: on the gendered and racialized exploitation of low-income service work in the "on-demand" economy', *Information, Communication & Society*, **20** (6), 898–914.

Van Slooten, J.M. and J. Holscher (2019), 'De werkerscoöperatie' ('The worker cooperative'), *Ondernemingsrecht*, **6** (1), 34–40.
Vieta, M., J. Quarter, R. Spear and A. Moskovskaya (2016), 'Participation in worker cooperatives', in D.H. Smith, R.A. Stebbins and J. Grotz (eds), *The Palgrave Handbook of Volunteering, Civic Participation, and Nonprofit Associations*, London: Palgrave Macmillan, pp. 436–53.
Webb, B.P. (1904), *The Co-Operative Movement in Great Britain*, London: Swan Sonnenschein.
Webster, J. (2016), 'Microworkers of the gig economy: separate and precarious', *New Labor Forum*, **25** (3), 56–64.
Whiting, M.E., D. Gamage, S.N.S. Gaikwad, A. Gilbee, S. Goyal, A. Ballav, et al. (2017), 'Crowd guilds: worker-led reputation and feedback on crowdsourcing platforms', in *Proceedings of the 2017 ACM Conference on Computer Supported Cooperative Work and Social Computing*, Portland, OR: ACM, pp. 1902–13.
Whyte, W.F. and K.K. Whyte (1991), *Making Mondragon: The Growth and Dynamics of the Worker Cooperative Complex*, 2nd edn, Ithaca, NY: ILR Press.
Wood, A.J., V. Lehdonvirta and M. Graham (2018), 'Workers of the Internet unite? Online freelancer organisation among remote gig economy workers in six Asian and African countries', *New Technology, Work and Employment*, **33** (2), 95–112.
Woodcock, J. and M. Graham (2020), *The Gig Economy: A Critical Introduction*, Cambridge: Polity Press.
Zamagni, V.N. (2012), 'Interpreting the roles and economic importance of cooperative enterprises in a historical perspective', *Journal of Entrepreneurial and Organizational Diversity*, **1** (1), 21–36.

10. The politics of platform work: representation in the age of digital labour

Paul Jonker-Hoffrén and Giedo Jansen

INTRODUCTION: THE PROBLEM OF REPRESENTATION

As shown in previous chapters, platform-based labour erodes traditional dichotomies between wage-employment and self-employment, and it increasingly decouples work not only from time (that is, work on demand), but also place. In this chapter we discuss how digital platform-based work challenges existing models of interest representation which are often based on the notions of place-based citizenship (that is, political representation) and/or type-of-contract segmentation (that is, representation in a country's systems of industrial relations). As regards the latter, falling in-between the category of wage-employment and self-employment, those working in the platform economy would not fit many of the existing systems of corporatist interest representation through which the interests of wage-employees tend to be represented by trade unions, and the interests of the self-employed are represented by branch, business and employer associations. The rise of platform work seems particularly challenging to trade unions. Since long-term, secure employment has dropped, trade unions in many countries face an erosion of their traditional membership base. At the same time, across Europe and other post-industrialised economies, trade unions have developed initiatives to give a voice to workers in the platform economy. Key examples are trade unions in, for instance, Belgium (Vandaele 2018), Sweden (Söderqvist 2018) or Switzerland that are attempting to conduct direct negotiations with the platforms (Vandaele et al. 2019).

In this chapter, we take political representation of digital gig workers to occur (potentially) in different spheres: the political arena (involving policy-making processes) and the industrial relations arena (involving worker representation). We are mostly concerned with substantive representation, that is, the degree to which representatives advocate the interests of a particular group. Descriptive representation, that is, the degree to which social-demographics of representatives mirror the characteristics of the represented, are largely ignored in this chapter. Substantive representation is a relational concept, linking differences in demand (in this instance, needs and preferences of gig workers) to differences in supply by (potential) representatives (in this instance, labour unions and political parties). We acknowledge that different countries have varying constellations of supply, and that in some countries the labour market relations system has a strong influence on the policy-making processes and vice versa. The aim of this chapter is to provide a generic theoretical framework representation in the age of platform-mediated gig work. We can only give some empirical examples to illustrate our contribution. Yet, this theory-driven view of gig worker representation should stimulate rich empirical studies of this largely uncovered topic, potentially engaging sociologists, political scientists and organisation scholars.

Research Gap: What We (Do Not) Know about Representation in the Gig Economy

The premise of much of the literature in sociology and political science, and related disciplines, has been that inequalities in the economy translate into patterns of interest representation. Labour market risks and job precariousness are increasingly identified as new lines of political division (Rehm 2009; Corbetta and Colloca 2013; Marx 2014). Although platform-enabled gig work has been around for some time now, there has been surprisingly little scholarly engagement with the topic in the context of interest representation. We identify four research gaps.

First, there is very little research on the individual-level preferences of platform workers (that is, the demand side of representation). As regards political preferences, recent literature has addressed the preferences of non-standard workers (for example, temporary employees, agency workers and the self-employed) vis-à-vis permanent employees (Corbetta and Colloca 2013; Marx 2014; Jansen 2016), but has not dealt specifically with platform workers. Turning to non-standard work more broadly, research has argued that workers who deviate from standard employment relationships often have distinct political preferences. Temporary workers are more supportive of redistribution policies and other social benefit programmes than are permanent employees, and more likely to support left-wing parties (Marx 2014). People in solo self-employment are more strongly orientated towards rightist positions regarding welfare policies and, in relation to party support, they more often prefer right-wing parties (Jansen 2016). Yet, also among the self-employed greater insecurities tend to spur stronger left-wing orientations, especially among high-skilled professionals (Jansen 2017). Falling in-between temporary work and self-employment, our understanding of the political preferences of platform workers is limited, especially given the heterogeneous composition of this group.

Second, equally unclear is the demand side of representation in the system of industrial relations. As regards trade unions, which is the most prominent form of employee representation, available individual-level studies yield inconclusive results. Newlands et al. (2018) found mixed opinions among platform workers regarding the need for a trade union. Among the largest platforms in their survey data, Newlands et al. (2018) found the most support for trade unions among Uber drivers, and the least support among BlaBlaCar drivers. In their study among Deliveroo riders in Belgium, Vandaele et al. (2019) found positive rather than negative views regarding trade unions. Although valuable, the current empirical evidence for the individual-level preferences is still scarce, fragmented and largely descriptive. There is little research looking into the drivers of trade union attitudes among platform workers, that is, characteristics of workers and working conditions that explain not only their views towards unions, but also their likelihood to join. Studies on trade union membership generally hold that the benefits of membership, relative to the costs, depend on the degree of attachment to the labour force, and that the benefits of membership decrease as workers are less strongly attached to their jobs (Jansen et al. 2017; Jansen and Lehr 2019). That platform work is often a secondary or part-time activity therefore complicates labour representation.

Third, and related to the previous research gap, the literature has so far almost exclusively focused on representation via trade unions but ignored preferences toward non-union forms of interest representation. In addition to grass-root organisations, discussed in Chapter 9 of this volume, there is a gap in our understanding of the supply-side of platform worker's representation in the systems of industrial relations and social dialogue. In particular, the rise of new freelancer's unions could be an alternative mode of organisation among digital freelancers who do not feel at home with traditional trade unions (Jansen 2020). Research has begun to describe, mostly qualitatively, recent initiatives of trade unions and other grass-root organisations to recruit platform workers (Johnston and Land-Kazlauskas 2018; Vandaele et al. 2019). Yet, a systematic, comparative investigation of the policy proposals regarding platform work, across countries and types of organisations, is lacking.

Fourth, similarly the supply side of political representation is poorly understood. That is, our knowledge on the stance of political parties on issues surrounding platform labour is virtually non-existent, even descriptively. Again, we may resort to the literature on non-standard work more broadly, but even that segment of literature still in its infancy (cf. Picot and Mendéndez 2019). In one of the few international comparative studies of party's electoral manifestos, Picot and Mendéndez (2019) found that the saliency (and criticism) of non-standard work follows a left–right distribution. However, they also observe difference within the left: while more traditional left-wing parties often address other labour concerns, issues around non-standard work are most vocally expressed by left-libertarian parties. These observations resonate with the individual-level findings that both temporary workers (Marx 2014; Häusermann et al. 2015) and self-employed (Jansen 2016; 2017) tend to support left-libertarian parties.

CAUSES FOR PATTERNS OF (AND PROBLEMS WITH) REPRESENTATION

Supply-side Factors

To understand the challenges to (political) representation of platform workers, we commence with theorising on the differences in supply. We start with three general theoretical views on non-standard employment and political party positions, building on the work of Picot and Mendéndez (2019), which we further specify regarding platform work.

First, traditional class-based theories would hold that left-wing parties represent the working class as a whole and strive for policies benefiting labour interests, such as generous social policies (Häusermann et al. 2013; Jansen et al. 2013). According to Picot and Mendéndez (2019) this view would lead to the expectation that left-wing parties represent and defend also the interests of non-standard workers. Extending this line of reasoning to platform work, it should mainly be left-wing parties, and by extension trade union, that address platform work most prominently, and express most concerns about this type of work.

Second, challenging the notion of a homogenous working class, insider–outsider theories (Doeringer and Piore 1971; Lindbeck and Snower 1989) assume that there is type-of-contract based segmentation in political appeal (Rueda 2005). In his 'insider-outsider model of partisanship', Rueda postulates that insiders are workers with highly protected jobs, whereas outsiders are 'either unemployed or hold jobs that are characterised by low salaries and low levels of protection, employment rights, benefits and social security privileges' (Rueda 2005, p. 62). Rueda posits that insiders favour policies geared toward employment protection and minimising wage competition, whereas the reverse is true for outsiders. Outsider would be concerned with their own job insecurity but not per se with protecting the secure jobs of insiders. The insider–outsider divide would mainly pose a dilemma to social democratic parties (Lindvall and Rueda 2014). Social democrats would represent the interests of insiders rather than outsiders (for example, Rueda 2005, 2007), since should they turn to outsider interests, they lose the insider and middle-class vote in the centre (Picot and Mendéndez 2019). Following this line of reasoning, there would be a representation gap on behalf of outsiders, and no party is expected to clearly address their interests. Again, what is expected to hold for outsiders in general, might hold for platform workers specifically. Being largely excluded from social protection, and with their hyper-flexible forms of work, platform workers are perhaps the 'ultimate outsiders' (Jansen and Lehr 2019). Extending the insider–outsider division to trade unions, unions would thus advocate policies that aim to protect insider jobs and wages at the cost of supporting legislation that allows outsiders, including platform workers, to improve their position by relaxing stringent employment protection. Trade unions would also be unlikely to support overly expensive universalist welfare state provisions aimed at improving the living standards for outsiders with unstable jobs. This prediction is backed by Jansen and Lehr (2019), who show that outsiders are less likely to be a union member and less willing to join trade unions than are insiders.

Third, challenging the notion of a representation gap, electoral realignment theories put forward that new political parties would be sensitive to the issues of non-standard workers. Picot and Mendéndez (2019) mention that left-libertarian and green parties would be an attractive alternative for those who are disappointed by social democratic parties. Unlike social democrats, these parties often have no, or weaker, ties with trade unions, which would otherwise link them to insider-orientated demands (such as dismissal protection). Moreover, these parties historically emphasise the needs of marginalised groups in capitalist societies. Also, given their orientation on minority and women's interests, they tend to favour universalist social policies, which are more accessible to non-standard workers with unstable employment careers. Based on a comparative investigation of party programmes in Germany, Italy and Spain, Picot and Mendéndez (2019) find most support for this third line of reasoning, that is, small left-wing parties are most vocal about non-standard work and that left-libertarian parties address the issue most specifically.

Demand-side Factors

Except for the traditional class-based theories assuming homogenous worker interests, most models of representation assume interest heterogeneity. Insider–outsider theories assume that workers with secure employment (that is, insiders) and those without (that is, outsiders) have distinct and opposing interests regarding which labour market policies would be in their favour. Other chapters in this book have already stated, however, that classifying platform work is not straightforward. This also applies to the binary insider–outsider distinction. There are at least two opposing views on the risk and returns of platform work. On the one hand, the negative view argues that platform labour erodes traditional employment relationships. The perception is that platform workers are misclassified as independent contractors or self-employed and are thereby excluded from labour law protection and from social security, leading to greater insecurity and poorer working conditions. From this perspective, platform workers belong to the group of outsiders. On the other hand, platform labour is primarily seen as a freelance phenomenon, contributing to employment flexibility, micro-entrepreneurship and new sources of income. In this perspective platform labour is based on the deliberate choice of engaging in digitally mediated work activities, often as a secondary source of income (Berg 2016). Especially when carried out in addition to a regular job, it is difficult to classify these forms of employment as outsiders. Instead, they may be considered part of a third category of 'upscale' groups. The upscales are those with a privileged position on the labour market, including 'self-employed professionals, shop owners and other business proprietors (Rueda 2007, p. 39). Upscale groups are distinct from insiders (who benefit from employment protection) in that they would support flexible arrangements, but they are also differ from in that they would oppose policies involving higher taxes to sustain generous social policies (Rueda 2007; Jansen 2017).

Building on Rueda's binary distinction, other scholars have proposed more fine-grained conceptualisations of outsiderness related to either more detailed classifications of employment status or the propensity of social risk (cf. Emmenegger 2009; Rehm 2009; Häusermann and Schwander 2011; Rovny and Rovny 2017). Extending this reasoning to platform labour, we suggest researchers should consider individual-level differences in employment risks when theorising on the political and policy preferences of platform workers. Differences in risks or returns of platform labour are likely to shape the demand for, diverging forms of, representation. Relevant individual-level characteristics are platform workers' socioeconomic backgrounds (occupation and education), work situations (for example, combining various sources of income) and motivations (for example, necessity driven versus opportunity driven) (Berg 2016), and more gig economy specific characteristics are also relevant, such as their online reputation (Wood et al. 2018). The core assumption is that different types of platform workers have different resources, opportunities and vulnerabilities. Necessity-driven platform workers working with greater insecurity and with no other sources of income are expected to behave more like outsiders, that is, turning away from social democratic parties and trade unions; whereas opportunity-driven self-reliant professionals are more likely to behave as entrepreneurs or upscales and support free-market right-wing parties, and associate with business or freelancer interest organisations (Jansen 2017; 2020).

Moreover, next to differences in the economic appeal of parties or organisations, prior individual-level studies have also examined non-economic factors driving the demand for more collectivistic forms of representation (for example, via trade unions, or left-wing politics). Most notably, the social context would matter, that is, the degree of social identification with co-workers, social cohesion at the workplace and feelings of solidarity. Recent research by Wood et al. (2018) and Newlands et al. (2018) suggests that variation among platform workers, to the extent that individuals (use online communication to) connect to other workers, may account for differences in trade union attitudes. At least three key characteristics of digital platforms are likely to affect the degree of trust in and identification among platform workers: (1) the degree of competition among providers, (2) the degree of algorithmic control over the work process, and (3) the degree of (de)localisation of services.

As regards the first of these, competition, many platforms employ forms of gamification in order to have platform workers compete against each other. Schreyer and Scharpe (2018) show how this works for Foodora. The desire or demand to be high in the rankings has explicit consequences for working conditions, as the work is physically demanding. Studies on traditional employment show that competition among workers undermines mutual interests (Storey and Bacon 1993; Wise 1993; Gunnigle et al. 1998) and weakens collective organisation (Jansen and Akkerman 2014; Jansen et al. 2017). Gamification of platform work nonetheless depends on what the work is; Foodora or Uber workers cannot in practice share their work. As Morschheuser et al. (2018) show, there is crowdsourcing work (as a subspecies of platform work) that can also be achieved in teams. This would lead to the expectation that the stronger the competition among providers, the stronger the individualist orientation, and the weaker the support for collective policies reducing employment risks. Yet, apart from the competitive element, gamification of work may not only be a threat to collective action. Loh (2019) is concerned with the gamification of political participation and argues that gamification can also lead to calls to action and viral discussions. Secondly is the key characteristic of algorithmic control. We expect that the openness of platforms to negotiate with their users affects the ability of workers to deal with insecurities. Strong algorithmic control, for example, over payment decisions, hinders the bargaining power of platform workers (Wood et al. 2019). The expectation would be that the more algorithms (instead of humans) decide how platform work should be allocated, performed and remunerated, the less able are workers to develop individual risk management strategies, and thus the more they are reliant on collective risk-reducing arrangements. The third characteristic is local versus delocalised services. Delocalised crowdworkers operate exclusively online through platforms that connect an infinite number of providers and clients, even across large geographic distances (Wood et al. 2018). Locally bound work on-demand via apps is platform-facilitated yet place-based work. An expectation would be that a lack of co-presence in delocalised work, makes it harder to develop a feeling of solidarity towards others (Lehdonvirta 2016; Graham et al. 2017) and, consequently, makes support for trade unions and left-wing, collective social security policies less likely.

CONSEQUENCES OF PATTERNS OF (AND PROBLEMS WITH) REPRESENTATION

The main consequences of representation of platform work lie in the sphere of social and employment policy. This chapter is not the right place for an in-depth review of policies or policy proposals in various countries but, notwithstanding cross-country differences, we argue that the main policy issues as regards platform-mediated gig work fit with at least two generic policy dimensions (see Figure 10.1). In Figure 10.1, we depict four ideal-types of policy solutions for platform workers regarding labour market risks. These are ideal-types that can be used to analyse political solutions that depend on national institutional configurations. Although these ideal-types work at a high level of abstraction, we argue that they cover the necessary dimensions.

The dimension of employment form (horizontal axis), between labour law and competition law as discussed in Chapter 4 in this volume, is in many countries the dividing line for access to income-dependent social security. This dimension is strictly related to what is legally possible in a jurisdiction, that is, whether a platform worker is seen as an employee or a self-employed entrepreneur. The dimension on collective/mandatory and individual/voluntary risk management (vertical axis) is to some extent based on the correct classification of workers and existing institutional arrangements, but arguably this is a dimension which can accommodate policy preferences ranging from statist to liberal.

Figure 10.1 thus shows four ideal-types: (A) mandatory-labour law, (B) mandatory-competition law, (C) voluntary-labour law and (D) voluntary-competition law. The latter, the voluntary-competition law ideal-type is closest to what in ordinary times is the case for most entrepreneurs regarding for example, pension savings. They are classified as entrepreneurs and are themselves responsible for arranging their pension savings. We name this ideal-type the conservative representation position, as it leans on existing interpretations of self-employment as a form of entrepreneurship and in that context connects to liberal values of individual responsibility. We expect that as a policy preference, this would fit to platform or self-employed workers with high incomes and specialised knowledge, such as information technology consultants. The voluntary-labour law ideal-type (type C) is almost an oxymoron, as it would envision platform workers (or self-employed workers)

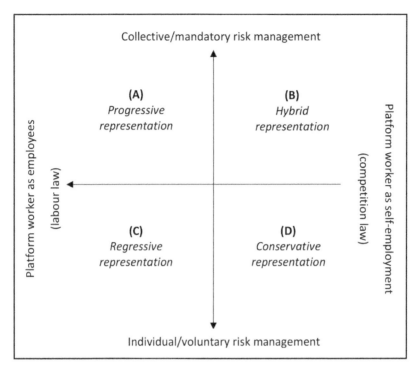

Source: Adapted from IBO (2015).

Figure 10.1   *Four ideal-types of representation positions regarding platform workers*

as employees while categorising them in a different position regarding social security. We name this ideal-type the regressive representation position, as it would do little to improve the position of platform workers. The target audience for this type of policy proposal is unclear, but it could be representative of a middle category, such as the US category dependent contractor (Bankhead and Petersen 2019). The mandatory-competition law ideal-type (type B) is interesting, in that it is a form of representation that acknowledges the special position of platform and self-employed workers as regards social risks. These risks are different from normal entrepreneurs. Therefore, we name this the hybrid representation position, because it applies the logic of employee social protection and social security to a special case of the legal category of entrepreneurs. This category could be the policy preference for those that argue that the position of dependent contractors should be improved. The mandatory-labour law ideal-type (type A), finally, is a form of representation that argues that platform workers or self-employed workers are sometimes employees and that, therefore, mandatory social security regulations should be valid for them too. We term this the progressive representation position. This position is in line with the extension of recent European Union (EU)-level legal judgements on the de facto position of self-employed workers (see Chapter 4 in this volume), and the recent initiative started by the European Commission to allow the self-employed to bargain collectively. Therefore, this policy position is pragmatic, but also prudent, and would suit the broad political centre.

Political Parties and Platform Work

Using the four ideal-types of representation positions in Figure 10.1, we return to the supply-side of political representation. There is virtually no extant literature, either theoretically or empirically, on party positions regarding platform work. Even the literature on party positions on non-standard work more broadly is scarce, as Picot and Mendéndez (2019) have shown. To provide a better understanding of the four ideal-types of representation positions, we conducted a small-scale inventarisation of policy positions of Dutch and Finnish political parties. Included are the four main parties in Finnish politics (Table 10.1) and all the parties that are currently in the lower house of the Dutch Parliament (Table 10.2). The first example of the political views on platform work (Table 10.1) is from Finland. Here, we used the 2019 election programmes of the four biggest parties. We searched for paragraphs, standpoints and discussions of the following keywords (translated from Finnish, and various spellings): self-employed, part-time, flex, flexible work, platform, temporary work, and permanent work. The party names are shown as the customary Finnish abbreviations. We coded the empirical material to enable comparison. In both tables we follow the coding of Picot and Menendez (2019), but we allow for more detailed information regarding policy proposals, which are indicated by the letter 'p' after the standpoint indicator. In addition, we mention if the parties have something to say about the role of social dialogue in relation to labour market issues.

For the Netherlands (Table 10.2), we studied the most recent election programmes of these parties using the same methodology (for the 2017 elections). The Freedom Party (PVV) and Forum for Democracy (FvD) are underrepresented in Table 10.2. These two parties had relatively short and generic election programmes, so their stance on these issues is unclear.

In the Finnish table (10.2), the Finn Party (PS) clearly states in their election programme, that their policy proposal is to improve the security of entrepreneurs. It mentions bankruptcy rules, social security and occupational health. Since the election programme mentions especially solo self-employed

Table 10.1   Positions of Finnish political parties in relation to non-standard work and platform work

	Flexwork	Self-employment	Platform work	Social security reform	Labour market structure reform	Tax system reform	Social dialogue	Representation position
PS	+	+ +	+	+ p	+ p	+ p	+ p	Hybrid
SDP		– p	– p	+ p			+ +	Hybrid
KOK	+ +			+	+ +	+ +		Hybrid
KESK	+	+		+ p	+ p	+ p	+	Progressive

Table 10.2   Positions of Dutch political parties in relation to non-standard work and platform work

	Flexwork	Self-employment	Platform work	Social security reform	Labour market structure reform	Tax system reform	Social dialogue	Representation position
VVD	+ +	+		+	+ p	+ p	– p	Inconclusive
PVV								No data
CDA	– –	– p		+ p	+	+ p	+	Inconclusive
D66	–	+		+ p	+ p	+ p	+	Inconclusive
GroenLinks	– –	– p		+ p	+ p	+ p	+ p	Inconclusive
SP	–	– – p		+ p	+ p	+ p		Inconclusive
PvdA	– –	– p	(+ p)	+ p	+ p	+ p	+ p	Progressive
ChristenUnie	–	+		+ p	+	+		Inconclusive
PvdD		o		+ p		+ p		Inconclusive
50Plus		+		+				Inconclusive
DENK		+		+		+		Inconclusive
SGP	–	+		+	+ +	+		Inconclusive
FvD						–		Inconclusive

and micro-entrepreneurs, it is evident that in this respect the Finn Party is in the hybrid representation position. The National Coalition Party (KOK) is not specific in their election programme about self-employed worker's social security, but they do mention their preferred social security reform ('general assistance') would be open also to entrepreneurs. The difference here is nonetheless that the National Coalition Party advocates social security reform that bases the reception of social assistance on particular conditionalities. The National Coalition Party would be in the same hybrid representation position, but since its focus is more on the social security dimension, it would be closer to the centre of the graph (Figure 10.1), especially since it advocates a complicated model in which freedom of choice and the private sector have a strong role. The Centre Party (KESK) can be located in the progressive representation position, as it makes a distinction between entrepreneurs and employees and advocates an improvement to their position. Their proposals regarding labour market relations imply that current legislation would also be extended to bogus self-employed workers. Finally, the Social Democratic Party (SDP) advocates a general security model similarly to the National Coalition Party, but without the obligatory aspect and a clearer collective dimension, since they state it is for everybody from entrepreneur to grant scholar. Therefore, it would be in the hybrid representation position, but further away from the centre of the graph (Figure 10.1).

Finnish politics is generally highly consensus based, and our interpretation of the policy positions of the four largest parties regarding platform/self-employed work shows this characteristic. It is important to remember that these are election statements. Without fully fledged policy proposals it is not possible to consider the precise forms of conditionality, universality and eligibility the parties prefer in specific situations. However, in the Finnish political arena an overhaul of the social security system has been a topic of intense debate for many years now. From this short analysis it is therefore evident that the political fault lines run across the topic of conditionality and the broader topic of basic income, and not so much on the issue of voluntarism or platform workers as entrepreneurs.

A similar analysis can be made for the Dutch example. Although in 2017 the Dutch political parties did not focus on platform work, the discussions about non-standard work more broadly can be analysed using Figure 10.1. The liberal party (VVD) is the only party that explicitly would like to see more flexwork in the Dutch labour market. Most other parties state that flexibilization has gone too far or is in other ways not desirable. D66 and the Christian Union state that self-employment should be promoted for the sake of employment, but even these are not completely positive about self-employment. Other parties stress the negative aspects, the lack of access to social security or the lack of opportunities for employment contracts. All parties, apart from the PVV and FvD, state that there must be social security reform. The extent of the reform varies, but centres on collective pensions and occupational disability insurance. The liberals want reform too, but towards privatisation. All parties have something to say about labour market structure reform. This ranges from the liberals, who want changes in employment contract law, to the CDA and some centre-left parties, who want to implement policies that encourage permanent employment contracts. This category is mainly concerned with issues that relate to the balance between permanent and flexible employment. The category of tax reform is broad, but often relates to the tax exemptions and deductions that self-employed workers have, which some parties see as unfair competition. Furthermore, many parties want to ease the tax burden on labour, in order to make regular employment contracts more attractive. Finally, the role of social dialogue is mentioned by some parties. The liberals would like to abolish extension of collective agreements, but other parties, such as the Green Left, wish to see a strengthening of the local work councils. Finally, although no party explicitly mentioned platform work in their 2017 election manifesto, the issue seemed to have gained prominence since that time. Parties at the left of the political spectrum pay attention to platform work on their websites (cf. GroenLinks,[1] SP,[2] PvdA[3] and, to a lesser extent, PvdD[4]). The Labour Party (PvdA) has published the most extensive set of policy proposals, including policies reforming classification of platform work, adapting competition law to grant platform workers collective bargaining rights, granting platform workers consultation or co-determination rights concerning algorithmic management. It is striking that platform work is now mentioned by at least some of the Finnish parties although these programmes date from only two years after the Dutch programmes.

SOLUTIONS TO PROBLEMS OF REPRESENTATION

The supply of representation for digital platform work is, on the one hand, restricted by legal issues, such as the limits of wage setting in the context of competition law (see Chapter 4 in this volume). On the other hand, there is no a priori reason why there could not be actors beyond that limitation. As Meardi et al. (2019) show, there are many reasons to problematise the concept of representation, especially in connection with disadvantaged workers. They argue that, in the literature, representation is commonly placed in the context of legitimacy, levels of representation and efficiency. However, as Meardi et al. (2019) show, this type of lens presupposes representation operating in

---

[1]   GroenLinks: https://groenlinks.nl/nieuws/eu-regels-moeten-einde-maken-aan-schijnconstructies-platformwerkers (accessed 13 April 2021).
[2]   SP: https://www.sp.nl/nieuws/2018/02/van-kent-kabinet-moet-sneller-optreden-tegen-deliveroo (accessed 13 April 2021).
[3]   PvdA: https://www.pvda.nl/nieuws/herovering-van-de-platformeconomie/ (accessed 13 April 2021).
[4]   PvdD: https://www.partijvoordedieren.nl/standpunten/zzp-ers (accessed 13 April 2021).

a specific legal context, for example, employee representation. To go beyond representation that is locked to a specific policy context, it should be seen as a dialectical process, incorporating more political than legal issues. Therefore, research into supply-side factors should focus more on claims-making than on how existing representative actors take into account new disadvantaged groups.

Meardi et al. (2019) discuss three different arguments to claims-making as a way of representation. The first is deeper roots, in which the actor claims deep knowledge and sympathy with a specific group. These may be actors that claim to be authentic because they are similar to those that they represent. Thus, representative claims may follow from descriptive representation. Second, an argument may be of expertise and special credentials. Labour unions can use this argument, when they have been politically and socially recognised as experts in the field (for example, on regulation or wage bargaining). The third argument is what is termed the 'wider interests and new voices' argument, which can especially be used in pluralistic labour market systems. These new voices are not left-wing per se, and can be politically to the right. Similar to the arguments of Picot and Mendéndez (2019), the new voices may arise when who can best represent outsiders is contested. As Meardi et al. (2019) note, there must be a constituency and an audience for the claims of representation. This is intricately linked to the preferences of the potential constituency: is representation a way to be heard in the political process or is representation a way to gain access to services?

Any potential supply-side solution to the representation problem must be sensitive to the legal context. When platform workers are not always clearly employees, it is not set in stone that formal instruments of representation, such as collective agreements, are the appropriate solution (see also Wright et al. 2017). Labour market instruments such as these may at first appear to be the logical step (incorporating outsiders in the insider framework) but this may not always be desirable, perhaps for legal reasons or owing to the preferences of the represented. Instead, representation should acknowledge the real work situation of the platform workers, in order to be able to offer the representation they need. This, however, demands that the preferences of platform workers, or the work circumstances, are known. One issue is the bond between platform worker and the platform: is it the only source of work? What is the duration of the relationship with the platform?

One way forward, despite the potential legal obstacles, is diversity management in unions (Greene et al. 2005). Diversity management signifies at least three issues: a neutral description of variability in a given workforce, a policy approach and a theoretical approach based on equality in diversity (Liff 1997). This demands that, within unions, there should be experts on platform work. In this instance, representation of these outsiders again becomes an internal issue for unions. However, given the current scale of platform work in combination with declining union membership, it seems less likely that general unions will pursue this issue. However, occupational or sectoral unions may be better positioned to do so, as their members' work may in the future be undertaken through platforms.

Beyond the organisation-based diversity management in unions and enterprises is the idea of worker centres. Worker centres are community-based organisations that fill the gap left by de-unionisation and retreat of government institution, often acting as quasi-unions (Martin et al. 2007; Theodore et al. 2009). Worker centres often work on behalf of migrants and other vulnerable groups. Sullivan (2010) argues that worker centres can constitute a revitalising factor for unions, since these community-based actors are not bound by legal constraints in the same way as normal unions are. In a more critical analysis, Visser et al. (2016) argue that worker centres do have an important function to integrate vulnerable workers into the main labour market, but owing to their institutional position, they will always be limited in their capabilities. A community-based worker centre has a physical presence, typically close to the hiring site (Theodore et al. 2009). The digital nature of platform work makes this aspect very problematic, although it could be imagined that for, for example, Foodora-riders, a worker centre would be located in an area with a high density of restaurants. Although a worker centre for platform workers cannot be literally close to the hiring site, the platform work that is physical as opposed to virtual could be supported by worker centres. Also, worker centres could have a virtual dimension, as Brady et al. (2015) show that digital advocacy can be an effective tool for community organising.

DIRECTIONS FOR FURTHER RESEARCH ON REPRESENTATION OF GIG WORKERS

The key to understanding representation of platform work is in comparative, multidisciplinary research. This chapter has outlined the sources, consequences and possible solutions to the representation gap of platform workers. We have argued that a large part of this gap results from, on the one hand, the unclear employment status of platform workers and, on the other, the outsider status that is a consequence of the unclear employment status.

One line of research of platform work, in relation to representation issues, is to study the potential to regulate the platform economy. As Stewart and Stanford (2017) show for Australia, there are several options for regulators, which mostly depend on how to view the employment relationship, the platform worker and the platform (as an employer). Nonetheless, regulation demands that platform work is compared with other forms of work (such as zero-hour contracts and self-employment). Without this comparison, it would

be difficult to legitimise either similar or different legal standards of regulation. Furthermore, regulation also implies that politics has ideas about how
to represent the relevant stakeholders in the policy-making process. Platform
firms may have the wherewithal to influence the process, but how is the view
of platform workers included (Collier et al. 2018)?

The most urgent line of research relates to the views of platform workers
themselves. Currently, there is insufficient knowledge relating to if, and to
what extent, they identity as a group with a shared interest, which may be
a prerequisite for collective action. Any study on this topic should acknowledge the variety of platforms and people working via platforms. For those
who gain the majority of their income from a platform, this type of work may
be far more central to their social identity (and thus in shaping their material
needs and political preferences) compared with those who casually work on
a platform. These issues should be studied holistically: not just the work, but
also the platform workers' position in relation to pension schemes, health-care
availability, social security hurdles; that is, how platform work relates to their
status as citizen and taxpayer.

In addition to their opinions, the position of platform workers as a form
of intersectionality should urgently be studied further. Platform workers
may not only be identified as platform workers (it remains to be seen if they
self-identify as these at all), but they also have other societal positions as
women or ethnic minorities. Studies by, for example, Van Doorn (2017) and
Adams and Berg (2017), show how the intersectionality of platform work and
gender and/or ethnicity provide a focal point for the analysis of these platform
workers' labour market position. For some women and ethnic minorities,
platform work may be one of the few pathways to labour market participation.
As regards union representation, however, both women and minority groups
remain underrepresented. Intersectionality between platform work, gender and
ethnicity may therefore amplify this marginalisation and widen the representation gap.

We do not know a great deal about the views political parties hold on platform work. The condensed empirical material in this chapter offers a glimpse
of the views of selected parties in a few countries. It is possible that, as the
platform economy politicises, political parties will produce clearer visions on
platform work in general and its consequences for, for example, social security
and taxation. Beyond these issues, it should be studied how political parties
would regulate platform work.

Collective action is dealt with in Chapter 9 of this volume, but issues of representation remain important to labour unions. Further research should focus
on ways in which labour unions can reach platform workers in many countries
where unionisation rates have been declining for many years, one reason for
which is that unions often represent the insiders rather than (or as well as) the

outsiders. Also, platform workers with a sufficiently high income may see unions as superfluous to their position (cf. Waddington et al. 2019, pp. 19–20). Representation of a heterogeneous group of workers will remain difficult for unions. Therefore, in-depth case studies of unions engaging with a variety of non-standard workers continues to be necessary. Non-union collective action should also be studied. Meardi et al. (2019) show there are various arguments for making claims of representation, and these can be made by non-union actors too. An interesting line of research may be in digital ethnography: how platform workers (or groups representing them) use social media to seek broader social attention.

Finally, policy-making at the EU level, which we hardly mentioned in this chapter, is a necessary route of inquiry, not only because platforms of digital work are the ultimate free labour market, but also because, at the EU level, employee representation is an even more difficult issue than at the national level. Dølvik (1997) discusses the logic of membership versus the logic of influence. Related to this supranational level is the notion that a great deal of platform work is or can be virtual, and therefore outside the polity. This is a complicated issue which would be interesting to study in the space between social-economic geography, economic sociology and international law, since this may also have implications for taxation.

ACKNOWLEDGEMENT

Paul Jonker-Hoffrén wishes to thank the Academy of Finland for the post-doctoral grant (grant number #307925) which enabled this study and his researcher mobility to the University of Twente.

REFERENCES

Adams, A. and J. Berg (2017), 'When home affects pay: an analysis of the gender pay gap among crowdworkers', accessed 15 April 2021 at https://papers.ssrn.com/sol3/papers.cfm?abstract_id=3048711.

Bankhead, S. and D.T. Petersen (2019), 'Finding the middle ground: establishing a third, hybrid worker classification', *Brigham Young University Prelaw Review*, **33** (1), art. 6.

Berg, J. (2016), 'Income security in the on-demand economy: findings and policy lessons from a survey of crowdworkers', *Comparative Labor Law and Policy Journal*, **37** (3), 543–76.

Brady, S.R., J.A. Young and D.A. McLeod (2015), 'Utilizing digital advocacy in community organizing: lessons learned from organizing in virtual spaces to promote worker rights and economic justice', *Journal of Community Practice*, **23** (2), 255–73.

Collier, R.B., V.B. Dubal and C.L. Carter (2018), 'Disrupting regulation, regulating disruption: the politics of Uber in the United States', *Perspectives on Politics*, **16** (4), 919–37.

Conen, W. and J. Schippers (eds) (2019), *Self-Employment as Precarious Work: A European Perspective*, Cheltenham, UK and Northampton, MA, USA: Edward Elgar.

Corbetta, P. and P. Colloca. (2013), 'Job precariousness and political orientations: the case of Italy', *South European Society and Politics*, **18** (3), 333–54.

Doeringer, P. and M. Piore (1971), *Internal Labor Markets and Manpower Analysis*, Lexington, MA: Heath.

Dølvik, J.-E. (1997), *Redrawing Boundaries of Solidarity?* Oslo: Arena/FAFO.

Emmenegger, P. (2009), *Regulatory Social Policy: The Politics of Job Security Regulations*, Berne: Haupt Verlag.

Graham, M., I. Hjorth and V. Lehdonvirta (2017), 'Digital labour and development: impacts of global digital labour platforms and the gig economy on worker livelihoods', *Transfer: European Review of Labour and Research*, **23** (2), 135–62.

Greene, A.M., G. Kirton and J. Wrench (2005), 'Trade union perspectives on diversity management: a comparison of the UK and Denmark', *European Journal of Industrial Relations*, **11** (2), 179–96.

Gunnigle, P., T. Turner and D. d'Art (1998), 'Counterpoising collectivism: performance-related pay and industrial relations in greenfield sites', *British Journal of Industrial Relations*, **36** (4), 565–79.

Häusermann, S. and H. Schwander (2011), 'Who are the outsiders and what do they want? Welfare preferences in dualized societies', *Les Cahiers Européens de Sciences Po*, **1** (July), 1–31, accessed 13 April 2021 at https://www.sciencespo.fr/centre-etudes-europeennes/sites/sciencespo.fr.centre-etudes-europeennes/files/01_2011%20who-are-the-outsiders-and-what-do-they-want-welfare-state.pdf.

Häusermann, S., T. Kurer and H. Schwander (2015), 'High-skilled outsiders? Labor market vulnerability, education and welfare state preferences', *Socio-Economic Review*, **13** (2), 235-258.

Häusermann, S., G. Picot and D. Geering (2013), 'Rethinking party politics and the welfare state – recent advances in the literature', *British Journal of Political Science*, **43** (1), 221–40

Jansen, G. (2016), 'Self-employment as atypical or autonomous work: diverging effects on political orientations', *Socio-Economic Review*, **17** (2), 381–407.

Jansen, G. (2017), 'Farewell to the rightist self-employed? "New self-employment" and political alignments', *Acta Politica*, **52** (3), 306–338.

Jansen, G. (2020), 'Solo self-employment and membership of interest organizations in the Netherlands: economic, social and political determinants', *Economic and Industrial Democracy*, **41** (3), 512–39.

Jansen, G. and A. Akkerman (2014), 'The collapse of collective action? Employment flexibility, union membership and strikes in European companies', in M. Hauptmeier and M. Vidal (eds), *The Comparative Political Economy of Work*, London: Palgrave Macmillan, pp. 186–207.

Jansen, G. and A. Lehr (2019), 'On the outside looking in? A micro-level analysis of insiders' and outsiders' trade union membership', *Economic and Industrial Democracy*, accessed 13 April 2021 at https://doi.org/10.1177/0143831X19890130.

Jansen, G., A. Akkerman and K. Vandaele (2017), 'Undermining mobilization? The effect of job flexibility and job instability on the willingness to strike', *Economic and Industrial Democracy*, **38** (1), 99–117.

Jansen, G., G. Evans and N.D. De Graaf (2013), 'Class voting and Left–Right party positions: a comparative study of 15 Western democracies, 1960–2005', *Social Science Research*, **42** (2), 376–400.

Johnston, H. and C. Land-Kazlauskas (2018), *Organizing On-demand: Representation, Voice, and Collective Bargaining in the Gig Economy*, Geneva: ILO.

Lehdonvirta, V. (2016), 'Algorithms that divide and unite: delocalisation, identity and collective action in "Microwork"', in J. Flecker (ed.), *Space, Place and Global Digital Work*, London: Palgrave Macmillan, pp. 53–80.

Liff, S. (1997), 'Two routes to managing diversity: individual differences or social group characteristics', *Employee Relations*, **19** (1), 11–26.

Lindbeck, A. and D.J. Snower (1989), *The Insider-Outsider Theory of Employment and Unemployment*, Cambridge, MA: MIT Press.

Lindvall, J. and D. Rueda (2014), 'The insider-outsider dilemma', *British Journal of Political Science*, **44** (2), 460–75.

Loh, W. (2019), 'The gamification of political participation', *Moral Philosophy and Politics*, **6** (2), 261–80.

Marx, P. (2014), 'Labour market risks and political preferences: the case of temporary employment', *European Journal of Political Research*, **53** (1), 136–59.

Martin, N., S. Morales and N. Theodore (2007), 'Migrant worker centers: contending with downgrading in the low-wage labor market', *GeoJournal*, **68** (2–3), 155–65.

Meardi, G., M. Simms and D. Adam (2019), 'Trade unions and precariat in Europe: representative claims', *European Journal of Industrial Relations*, **27** (1), 41–58.

Morschheuser, B., J. Hamari and A. Maedche (2018), 'Cooperation or competition - when do people contribute more? A field experiment on gamification of crowdsourcing', *International Journal of Human-Computer Studies*, **127** (July), 7–24, accessed 13 April 2021 at https://doi.org/10.1016/j.ijhcs.2018.10.001.

Newlands, G., C. Lutz and C. Fieseler (2018), 'Collective action and provider classification in the sharing economy', *New Technology, Work and Employment*, **33** (3), 250–67.

Picot, G. and I. Menéndez (2019), 'Political parties and non-standard employment: an analysis of France, Germany, Italy and Spain', *Socio-Economic Review*, **17** (4), 899–919.

Rehm, P. (2009), 'Risks and redistribution an individual-level analysis', *Comparative Political Studies*, **42** (7), 855–81.

Rovny, A.E. and J. Rovny (2017), 'Outsiders at the ballot box: operationalizations and political consequences of the insider–outsider dualism', *Socio-Economic Review*, **15** (1), 161–85.

Rueda, D. (2005), 'Insider–outsider politics in industrialized democracies: the challenge to social democratic parties', *American Political Science Review*, **99** (1), 61–74.

Rueda, D. (2007), *Social Democracy Inside Out: Partisanship and Labor Market Policy in Advanced Industrialized Democracies*, Oxford: Oxford University Press.

Schreyer, J. and J.-F. Scharpe (2018), 'Algorithmische Arbeitskoordination in der plattformbasierten Gig Economy: das Beispiel Foodora', *AIS-Studien*, **11** (2), 262–78.

Söderqvist F. (2018), 'Sweden: will history lead the way in the age of robots and platforms?', in M. Neufeind, J. O'Reilly and F. Ranft (eds), *Work in the Digital Age. Challenges of the Fourth Industrial Revolution*, Oxford: Rowman & Littlefield, pp. 295–304.

Stewart, A. and J. Stanford (2017), 'Regulating work in the gig economy: what are the options?', *Economic and Labour Relations Review*, **28** (3), 420–37.

Storey, J. and N. Bacon (1993), 'Individualism and collectivism: into the 1990s', *International Journal of Human Resource Management*, **4** (3), 665–84.

Sullivan, R. (2010), 'Organizing workers in the space between unions: union-centric labor revitalization and the role of community-based organizations', *Critical Sociology*, **36** (6), 793–819.

Theodore, N., A. Valenzuela and E. Meléndez (2009), 'Worker centers: defending labor standards for migrant workers in the informal economy', *International Journal of Manpower*, **30** (5), 422–36.

Vandaele, K. (2018), 'Will trade unions survive in the platform economy? Emerging patterns of platform workers' collective voice and representation in Europe', ETUI Research Paper – Working Paper, accessed 13 April 2021 at https://papers.ssrn.com/sol3/papers.cfm?abstract_id=3198546.

Vandaele, K., A. Piasna and J. Drahokoupil (2019), '"Algorithm breakers" are not a different "species": attitudes towards trade unions of Deliveroo riders in Belgium', ETUI Research Paper – Working Paper, accessed 13 April 2021 at https://papers.ssrn.com/sol3/papers.cfm?abstract_id=3402899.

Van Doorn, N. (2017), 'Platform labor: on the gendered and racialized exploitation of low-income service work in the "on-demand" economy', *Information, Communication & Society*, **20** (6), 898–914.

Visser, M.A., N. Theodore, E.J. Melendez and A. Valenzuela (2016), 'From economic integration to socioeconomic inclusion: day labor worker centers as social intermediaries', *Urban Geography*, **38** (2), 243–65.

Waddington, J., T. Müller and K. Vandaele (2019), 'Setting the scene: collective bargaining under neoliberalism. Collective bargaining in Europe: towards an endgame', in T. Schulten, J. Waddington and K. Vandaelde (eds), *Collective Bargaining in Europe: Towards an Endgame*, 4 vols, vol. 1, Brussels: ETUI, pp. 1–32.

Wise, L.R. (1993), 'Whither solidarity? Transitions in Swedish public sector pay policy', *British Journal of Industrial Relations*, **31** (1), 75–95.

Wood, A.J., M. Graham, V. Lehdonvirta and I. Hjorth (2019), 'Good gig, bad gig: autonomy and algorithmic control in the global gig economy', *Work, Employment and Society*, **33** (1), 56–75.

Wood, A.J., V. Lehdonvirta and M. Graham (2018), 'Workers of the Internet unite? Online freelancer organisation among remote gig economy workers in six Asian and African countries', *New Technology, Work and Employment*, **33** (2), 95–112.

Wright, C.F., N. Wailes, G.J. Bamber and R.D. Lansbury (2017), 'Beyond national systems, towards a "gig economy"? A research agenda for international and comparative employment relations', *Employee Responsibilities and Rights Journal*, **29** (4), 247–57.

11. Conclusion: solutions to platform economy puzzles and avenues for future research

Giedo Jansen, Victoria Daskalova and Jeroen Meijerink

INSIGHTS ON PLATFORM ECONOMY PUZZLES: CUTTING ACROSS THE CHAPTERS

The aim of this book was to offer a multidisciplinary perspective on platform-enabled gig work and the implications (both intended and unintended) this has for society, labour markets, public policy organizations and individuals. To realize this aim, each chapter in this edited volume examined a platform economy puzzle from the perspective of a given academic discipline. In this chapter we synthesize these insights in relation to the drivers, outcomes and solutions to different platform economy puzzles. We identify three overarching puzzles: (1) non-standard, multiparty working relationships and precarious working conditions; (2) the use of algorithmic management practices; and (3) the (excessive) capture of surplus value by platform firms. Although these three puzzles are tightly interrelated, we discuss them as stand-alone issues for the sake of analytical separation.

Non-standard, Multiparty Working Relationships and Precarious Working Conditions

Nature of the platform economy puzzle

The various chapters in the volume outlined how platform-enabled gig work is characterized by the absence of the standard employment relationship and the precarious working conditions that this creates for workers (Goods et al. 2019; Meijerink et al. 2019; Shapiro 2018; Van Doorn 2017; Wood et al. 2019). As discussed by Stanford (Chapter 3 in this volume), instead of representing a new type of work, platform work revives some of the challenges associated with older forms of freelance work that existed prior to World War II. The consequence of this resurgence of gig work is increased precarity for platform workers who miss out on benefits traditionally tied to the standard employment relationship, such as social security benefits, insurances and a stable income. Moreover, platform workers often lack social contact with peers, and instead have to engage in fierce competition with other labourers (across the globe) to gain access to work. In this respect, and as noted by Jonker-Hoffrén and Jansen (Chapter 10), platform-enabled gig workers can be seen as outsiders. Here, 'outsiders' refers to workers with low levels of protection, employment rights and social security privileges. As outsiders, it is likely that the working conditions of platform workers remain precarious, since politicians – social democrats in particular – are incentivized to avoid losing the vote of insiders, that is, the middle-class worker with stable employment contracts. Following this line of reasoning, it is precisely the lack of not having an employment relationship that makes it hard for platform workers to escape the precarity associated with them being outside the confines of an employment relationship.

Moreover, platform workers are a diverse group of individuals who differ in relation to their socio-economic background, interests and motivations. As noted by Jonker-Hoffrén and Jansen (Chapter 10), this creates challenges to collectivize platform workers and have labour unions represent their interests. Platform workers cannot expect much from consumers either. As shown by Smith et al. (2021), consumers have very little awareness of, or concern about, the precarious working conditions of platform workers. As discussed by Duggan et al. (Chapter 8), this creates additional problems in relation to psychological contract breaches when platform firms do not uphold the expectations of platform workers that platforms act against retaliating customers. As shown by Keegan and Meijerink (Chapter 7), platforms (algorithmically) delegate managerial activities (for example, appraisal, pay, instructions and job design) to consumers and hiring businesses. On the one hand, this enables platform firms to address institutional complexity in relation to tensions between worker autonomy and control. On the other, however, it creates problems for platform workers when consumers or hiring business do not live up to their responsibilities and thereby breach psychological contracts.

Solutions to the platform economy puzzle
The first solution to the precarious working conditions of platform workers is to re-establish the standard employment relationship and reclassify platform workers as employees. In the words of Jonker-Hoffrén and Jansen (Chapter 10), this would make platform workers 'insiders'. Turning platform workers into insiders is what labour unions attempt to achieve by initiating court cases on the false self-employment status of platform workers.

Second, precarious working conditions in online marketplaces can be (partially) addressed by entitling platform workers with the same protection and benefits traditionally associated with the standard employment relationship. This would require a change in labour law to ensure employment benefits are not exclusively tied to the employment relationship, as discussed by Stanford (Chapter 3) and Van Doorn and Badger (Chapter 6). Here, Jonker-Hoffrén and Jansen (Chapter 10) teach us that this requires collective, mandatory risk management (that is, social protection and security for independent workers or entrepreneurs) as well as individual, voluntary risk management (for example, opt-in for entrepreneurs to make pension savings). Ultimately, making social security benefits mandatory in both employment and freelance contexts has multiple benefits, as discussed by Keegan and Meijerink (Chapter 7); it reduces institutional complexity, alleviates tensions between worker autonomy and control, and obliges platform firms to offer additional worker benefits.

Third, precarious working conditions can be tackled by new forms of collective action as described by Van Doorn and Badger (Chapter 6) and Jonker-Hoffrén and Jansen (Chapter 10). This involves information sharing and building networks of platform workers to overcome social isolation. More advanced forms of collective action involve the creation of quasi-unions and work centres to cater to the needs of platform workers with different socio-economic backgrounds, interests and motivations. Here, the measurement of platform work prevalence, as outlined by Pesole (Chapter 2), is important to gauge the diversity in types of platform work(ers). Finally, Bunders (Chapter 9) proposes the creation of platform cooperatives. Platform cooperatives help in addressing precarious working conditions while their members (that is, platform workers) collectively make up for risks (for example, temporal lack of income and health insurances) that are traditionally covered by an employer. Also, platform cooperatives resolve institutional complexity associated with human resource management (HRM) without employment (Chapter 7) as worker-members of cooperatives can offer training to one another and engage in democratic decision-making, without facing reclassification issues.

The Use of Algorithmic Management Practices

Nature of the platform economy puzzle

Online labour platforms rely on a range of software algorithms to augment and/or automate managerial decision-making in areas such as hiring and firing, performance appraisal, pay and task allocation (Jarrahi and Sutherland 2019; Leicht-Deobald et al. 2019; Newlands 2020). This creates a number of challenges. For instance, and in line with the work of Duggan et al. (Chapter 8), algorithmic management may breach psychological contracts – especially when software algorithms are designed to be self-learning and adaptive. For instance, on a given day, algorithmic management practices may build worker expectations about (future) pay, task allocation or feedback, while on another day, they may violate expectations when algorithmic decision-making processes have changed. This creates high levels of uncertainty on the part of platform workers. Moreover, the use of algorithmic management obfuscates the (employment) relationship between workers and platforms. As noted by Keegan and Meijerink (Chapter 7), platform organizations implement HRM without employment, by means of software algorithms that exercise control over workers. It is the opacity of HRM algorithms that obfuscates the control that platforms exercise over workers, adding to the puzzle of whether platform workers should be seen as freelancers or employees. Ultimately, as discussed by Van Doorn and Badger (Chapter 6) as well as Duggan et al. (Chapter 8), this provides platform firms the opportunity to exploit platform workers through automated wage theft, discriminating between workers and monetizing the data generated by workers.

Solutions to the platform economy puzzle

Although the chapters in this volume offer limited solutions to the challenges related to the use of algorithmic management, some insights can be derived from the literature. These solutions include the creation of algorithmic accountability schemes, which requires online labour platforms to be transparent about their algorithmic decision-making processes (Ananny and Crawford 2018), allowing workers to build algorithmic competences (Jarrahi and Sutherland 2019) or hiring algorithmists to sharpen human oversight in algorithmic decision-making processes (Leicht-Deobald et al. 2019).

Algorithmic management can also be turned into a solution to some of the puzzles outlined in this volume. For instance, owing to their potential to collect, process and analyse data at scale, software algorithms may support the measurement of the prevalence of gig work as discussed by Pesole (Chapter 2). This involves relying on data coming from smartphones to gauge the number of gigs performed by freelance meal deliverers, processing financial data collected by banks on the financial compensation of platform workers or reverse engineering the algorithms operated by platform firms. Moreover, algorithmic management may aid platform workers in setting up platform cooperatives. As noted by Bunders (Chapter 9), owing to a lack of financial resources, platform cooperatives need to limit operating costs. Here, algorithmic management practices can automate some of the managerial activities that coordinate the efforts of cooperative members and, ultimately, cut costs.

The (Excessive) Capture of Surplus Value by Platform Firms

Nature of the platform economy puzzle

Online labour platforms capture (excessive amounts of) surplus value from workers and other stakeholders. First, platform firms can increasingly monetize the labour efforts of platform workers. As noted by Stanford (Chapter 3) and Van Doorn and Badger (Chapter 6), there is an oversupply of platform workers (amplified by the financial crisis and the COVID-19 pandemic) which creates power imbalance between platform workers vis-à-vis platform firms. In addition, (illegal) migrant workers may be heavily dependent on platform work as they have no other option but to turn to online labour platforms for generating an income. This puts platform firms in a powerful position to reduce workers' remuneration for capture surplus value. Moreover, platforms benefit financially from working with freelancers since this dismisses them from the obligation to offer secondary benefits and social security to platform workers. Ultimately, individuals and/or society have to make up for this, which ultimately undermines the social welfare state.

Second, Van Doorn and Badger (Chapter 6) show how online labour platforms capture value from the data that workers generate (dual-value production). On the one hand, this involves platforms increasing profit margins by algorithmically processing worker-generated data for dynamically adjusting workers' remuneration. On the other, worker-generated data aids value capture while it has speculative value to future investors. This speculation manifests when investors expect that data-rich platform firm are able to create data-driven cost efficiencies, cross-industry synergies and new markets. This creates a promise (whether true or not) of near-future competitive advantages that motivate future investors to invest their money in platform firms. This allows early investors to cash out when filing for an initial public offering (IPO), thereby capturing value from the speculative data generated by platform workers.

Finally, in Chapter 5, Shapiro explains that online labour platforms capture infrastructural surplus, that is, the excess of value derived from collective resources embedded in the urbanised landscape. Among others, it involves platforms appropriating value from publicly available resources. Morcover, infrastructural surplus is indirectly captured by platforms through the knowledge and skills of their workers that need to navigate urban areas or when workers are expected to invest in the equipment needed to perform platform work (see also Chapter 3 by Stanford). The consequence of this is situations where urban actors, such as workers, municipalities and local shop owners, end up paying for the infrastructure from which platforms capture rent.

Solutions to the platform economy puzzle

A first solution, as suggested by Daskalova et al. (Chapter 4), is to allow freelance platform workers to collectively set prices charged to platforms. This requires a change in competition law which currently prevents freelance workers – as the smallest of small businesses – to bargain collectively. This change can be implemented in different ways, for example, by means of a special notification and exemption procedure, as used in Australia.

Second, as noted by Bunders (Chapter 9), platform cooperatives could restore the situation where investors capture excess value from workers' labour, while worker-members of platform cooperatives capture the profit they generate themselves. In addition, Van Doorn and Badger (Chapter 6) argue that the speculative value of worker-generated data would accrue to the workers generating the data as, ultimately, they are the owners of the platform cooperative they work for. Here, governments and labour unions can limit excessive value capture by platform firms, by subsidizing platform workers to set up a platform cooperative. Even if setting up a platform cooperative is not feasible, labour unions can assist workers in reclaiming the data that platforms use to exercise control over workers for reverse engineering algorithmic control processes of platform firms.

Finally, Van Doorn and Badger (Chapter 6) call for stricter tax regulations, such as taxes on revenues generated from worker-generated data. In addition, they propose limiting excessive value capture through the taxation of platform-related assets. This involves putting taxes on private equity transactions during IPOs. In turn, these taxes can be used to finance investments in public infrastructure as suggested by Shapiro (Chapter 5). More specifically, Shapiro advises public policy-makers to invest in affordable housing and healthcare. This ultimately avoids situations where residents have to complement their income by means of precarious gig work. Since platform firms benefit financially from resources in local public landscapes, Shapiro recommends putting stringent taxation on platform firms and using these taxes to invest in public infrastructures.

AVENUES FOR INTERDISCIPLINARY RESEARCH INTO PLATFORM WORK

The chapters in this volume show that we can already learn a lot by looking back, at the scholarship and theoretical frameworks developed within different disciplines. Each chapter studied platform-enabled gig work from a different and more or less distinct disciplinary perspective. As these chapters underlined, many of the puzzles we face in the context of platform work are a revival of well-known challenges. This familiarity can serve us well. Yet, there are some novel aspects of platform-enabled gig work which require fresh scholarly attention, ideally from a multidisciplinary perspective. Specifically, we see the following sets of questions that would benefit from a combination of disciplines to inform future research into platform-enabled gig work.

Old and New Questions about Data Justice

As mentioned, gig work generates valuable data which flows from the workers' smartphone or computer to the platform. This data can be a tremendous source of value. How to distribute the gains from this valuable information (and how to prevent its misuse) is an economic, legal, social as well as political question. Various chapters in this volume have shown that the platform economy raises questions both about equity (for example, whether or not the workers should be compensated for this data) and about fair competition on the respective market (among platforms, between platforms and 'traditional' providers in the sector, or between platforms and alternative providers, such as cooperatives).

A number of contributions in this volume have mentioned access to data as an essential issue. Data generation is arguably the explanation behind the vigorous investment in money-losing platform companies, as argued by Van Doorn and Badger (Chapter 6). The authors argue that access to data allows platform companies to topple traditional competitors by providing more efficient services. It enables disruption of related industries (for example, the expansion of Uber into food-delivery services), threatening to subject a number of sectors to the dominance of one platform. Importantly, troves of data can also be used with the ultimate goal of automating the workers away (as with Uber's expansion into self-driving cars). This raises the question: if data determines winners and losers in a market, how can access to it be guaranteed on fair terms?

There are some bottom-up initiatives, which intend to contest the platform company's monopoly on data by generating alternative data-sets. That is, the tools for dual-value creation can also be put to work for users. As Keegan and Meijerink (Chapter 7) discuss, HRM activities for controlling requesters can also be used to enable dual-value production by, for example, restaurants (in relation to Uber Eats and Deliveroo). That is, restaurants also generate data themselves and may share and sell that information. Other initiatives, such as the Workers Information Exchange, rely on privacy protections (such as the General Data Protection Regulation, GDPR, in Europe) in order to re-claim worker-generated data. Data protection rights are also used by workers to press for data portability, to challenge algorithmic decision-making (for example, deactivation from the platform) and to substantiate employment-related claims in court. However, to what extent are these individual rights sufficient to achieve data justice? The question arises as to whether data protection legislation is not already well behind that which market players, as well as individuals, need in order to guarantee data justice and fair competition.

Finally, access to data is also essential for future innovators and for state authorities. As noted by Pesole (Chapter 2), researchers face troubles in estimating the size and depth of the platform work phenomenon. Yet, data is essential in making policy decisions, for example, on social security and minimum wages. Currently, one of the main sources of data are workers themselves. The question arises as to how policy-makers' access to data can be facilitated so that the policy decisions taken are better substantiated.

Old and New Questions about the Cooperation-Competition Puzzle

Cooperation among people is necessary for survival; yet, our societies also assume that competition can be beneficial to progress. So how do we reconcile the two? When is cooperation justified and when is competition fair? This issue is at the core of Chapter 4, by Daskalova et al., which discusses the tension between collective labour rights (such as the right to bargain collectively) and competition law (in particular, the cartel prohibition). The authors indicate that collective bargaining can achieve important efficiencies and that there are exemption possibilities. However, the competition–cooperation conundrum extends beyond legal matters. Cooperatives, as noted by Bunders (Chapter 9), also have to overcome cooperation-related problems, such as free-riding by members and competition by workers. Curiously, and often neglected, similar collective action problems may also be faced by platform corporations when coordinating their policy preferences and interest representation (for example, via business and trade associations vis-à-vis trade unions and governments) with other companies. These questions are as much a matter of theorizing in (business) economics and political science as they are a matter of legal considerations. As discussed by Keegan and Meijerink (Chapter 7), platform firms have to balance diverging institutional logics leading to disparate policy preferences: corporation logic emphasizes workers being embedded in the organization and subject to its control, whereas market logic emphasizes competition.

At the platform level, competition versus cooperation questions surface as well. As noted by Bunders (Chapter 9), platform cooperatives might be limited in their ability to compete with corporations owing to disparities in relation to financing or access to data. Arguably, the same holds for traditional companies in disrupted sectors. Investor-subsidized platforms can afford to engage in predatory strategies (for example, offering services to the consumer at below-cost prices) in order to achieve dominance on the market. Technology-shrewdness and the ability to collect and exploit rich data-sets also limit the expansion of alternative competitors, such as cooperatives or counter-strategies by traditional players. The question is then whether the state of play permits alternative competitors to grow? In addition to cooperation with other market players, the issue of internal collaboration allows for new interdisciplinary cross-overs. What is, for example, the role of HRM activities in platform cooperatives? May HRM activities be used to build trust and a shared identity and reciprocity among a heterogeneous collective of cooperative-workers? Especially when implementing strong collaborative models, platform cooperatives may introduce new types of institutional complexity.

Old and New Questions about the Sustainability Question

Whether one is critical or excited about platform economy developments, the question of long-term consequences is in order. The contributions in this volume have also considered to what extent platforms can sustain their growth. Speculative financing sustains the platforms' disruptive march forward, but one must wonder how long can investors tolerate loss on investments. Consumers enjoy the cheaper services enabled by the advent of the gig economy, but governments and workers might soon decide they are no longer willing to subsidize these lower prices and correcting for platform externalities, as noted by Shapiro (Chapter 5) and Daskalova et al. (Chapter 4). Hence, the question is not so much whether the platform-mediated gig model of work is sustainable in the long run, but for whom and under what conditions. Again, research moving these issues forward is not exclusive to one discipline, and should be theorized on the actions various parties at various levels of analysis play. Apart from a legal, economic and political question, the issue of the long-term sustainability may also look into historical precedents, and take into account (changing) social norms and expectations, including 'psychological contracts' between workers and consumers. Currently, however, research on the long-term consequences platform-mediated gig work is hindered by limited data availability and other methodological challenges that are discussed in the next section.

METHODOLOGICAL CONSIDERATIONS FOR FUTURE RESEARCH

As discussed by Pesole (Chapter 2), the empirical investigation of platform labour is still in stages of infancy and measuring platform-mediated gig work is a daunting task. Moreover, platform work may contain 'devil in the details' requiring the collection of qualitative data. When it comes to methodological and data collection approaches to study platform labour, the chapters in this volume raise three overarching empirical challenges and possibilities.

Empirical Challenges and Opportunities Regarding Micro-level Data

To date, most quantitative survey-based studies were aimed at estimating the aggregate size of the platform economy, and main employment statistics (see Chapter 2 for a discussion). But the current state-of-the art leaves many issues unaccounted for. Without the ambition of presenting an exhaustive list, we discuss three shortcomings. First, the chapter by Pesole (Chapter 2) emphasizes the importance of within-person variability. The reference period that is chosen to measure platform labour activities matters. To unravel such temporal dynamics, future research should invest in longitudinal micro-level data on platform work. For example, following the logic of the psychological contract in the chapter by Duggan and colleagues (Chapter 8), the time horizon should matter for multi-party relationships in the gig economy. The nature of an exchange and mutual expectations should weigh heavier under long-lasting and repeated interaction. Equally, changes in the HR management of platform workers and may affect how they experience work. Also, we do know little about how platform work fits the wider career patterns of individuals. Besides large-scale survey studies, this also calls for research designs such as ethnographic research, case studies, and naturalistic observations for capturing (micro-) changes in the working conditions and lived experiences of platform workers.

Second, it is common for platform economy research to study platform users (most often workers or consumers). Far less attention is paid to non-users, potential users, or to people who are affected by platform economy, but who do not provide services themselves. Hence, relevant comparison groups are needed. The legal chapter by Daskalova and colleagues (Chapter 4) shows that platform work often falls in-between the category of (temporary) wage-employment and self-employment. Therefore, it makes sense to take the experiences and behaviour of these two groups as a benchmark to study the experiences and behaviours of platform workers. Bunders (Chapter 9) for example suggests to use traditional forms of collective action (for example by employees in trade unions) as a comparison to understand worker-owned platform cooperatives.

Third, besides data on employment activity and working conditions, only few researchers so far have investigated the attitudes and behaviours of platform workers. This behaviour should primarily involve the tasks that are carried out by various platform users/actors. Pesole (Chapter 2) points to the importance of a task-based approach to understand thoroughly how work content and work organization are affected by platforms' intervention. But the emphasis on attitudes and behaviour should also be extended to aspects of social and political lives that are affected by people's work status and experiences. For instance, Jonker-Hoffrén and Jansen (Chapter 10) discuss relationship between gig work and voting behaviour/union membership. Yet, future research may explore other effects of gig work beyond the workplace, for example in the area of work-life spill-overs and social cohesion. On the one hand – echoing the critical perspective – gig work may have an isolating effect, reducing platform workers involvement in the wider community. On the other hand, increasing flexibility, and homebased work likely improves the work-life balance of workers, and removes constraints to active social/civic involvement.

Empirical Challenges and Opportunities Regarding Meso-level Data

Either implicitly or explicitly, several chapter in this volume call for a relational approach to platform labour. Duggan et al. (Chapter 8) most explicitly address the importance of the social ties between platform firms, gig workers, customers and suppliers. The quantitative study of relational ties requires dyadic data (for example, worker–customer interaction), triadic data (for example, workers, platforms and customer interaction) or network data (for example, interaction among workers of the same platform).

A relational approach is key for the chapters that look into collective organization (Bunders, Chapter 9) and interest representation (Jonker-Hoffrén and Jansen, Chapter 10). Owing to its episodic nature, platform-mediated gig work may diminish long-term co-worker interaction and thus reduce workers' perceptions of collective interests. However, people do not make the decision to participate in collective action in a social vacuum. Research by Wood et al. (2019) and Newlands et al. (2018) suggests that variation in the extent to which platform workers (use online communication to) connect to other workers, may account for differences in collective organization. It is therefore worth adopting longitudinal research designs that tap into platform worker interactions.

Empirical Challenges and Opportunities Regarding Macro-level Data

Future research should supplement individual-level data with country-level variables, and use cross-level interaction models to assess how the macro-level context affects the attitudes and behaviour of platform workers. The work status of platform worker is strongly dependent on the national legal framework. The legal-institutional context therefore conditions the risks and benefits associated with platform labour. Yet, currently, we know little about how the institutional settings affect individual-level cost–benefit calculations of platforms when making choices between various platforms and/or activities. We know even less about the impact of country-level conditions that affect the wider social behaviour of platform workers. The chapters by Jonker-Hoffrén and Jansen (Chapter 10) and by Bunders (Chapter 9) indicate the importance of cross-country comparisons in studying political behaviour and collective action. Bunders argues how variation in laws and policies between countries could influence individual gig workers' demand for collective action and the opportunities and pitfalls for collective action in the platform economy. This resonates with Jonker-Hoffrén and Jansen who argue that the individual-level choices of platform workers to join a trade union or to vote for a particular party is not only driven by the variation in demand, but also by variation in (political) supply. Both chapters observe a lack of available data that enables the testing of these macro–micro interactions. The need to combine individual-level data with contextual characteristics is not limited to cross-country comparisons but, as Stanford (Chapter 3) shows, also holds for comparisons over time.

Here, we foresee a role for both supra-national bodies, such as the European Union, as well as meta-platforms as discussed by Van Doorn and Badger (Chapter 6). For instance, the European Union is well positioned to (further) develop data repositories on working conditions of platform workers across nations and how these are impacted by cross-country, institutional differences (for example, Eurofound's platform economy repository). Moreover, and with the help of GDPR regulation, (collectives of) platform workers and representative bodies (for example, the International Labour Organization, the European Trade Union Institute and the Organisation for Economic Co-operation and Development) may require meta-platforms to share worker-generated data for comparative cross-national studies into platform work.

CONCLUDING REMARKS

In this chapter we aimed to synthesize the insights from the various contributions in this volume. A few reservations apply. First, the three overarching problems identified in this chapter are by no means exhaustive. Readers will find many more similarities and connections across chapters. The same holds for our suggestions for further research. Second, the synthesis presented here follows from our interpretation, as editors, of the contributions from the individual authors. Hence, any misinterpretations or simplifications regarding the original arguments are our own. Finally, most of the research presented in the preceding pages was completed prior to the onset of the COVID-19 pandemic. The crisis has surfaced and sometimes amplified some of the problems discussed in this volume, and emphasized both the need for flexibility and the need for social protection of platform-mediated gig work during times of crisis. Yet, not all of these developments could be integrated in this book. Despite these limitations, we are confident that our book presents the state-of-the-art in theory and research on platform-mediated gig work. We hope that this volume inspires not only researchers, but also policy-makers and practitioners to take up existing and future challenges of the platform economy.

REFERENCES

Ananny, M. and K. Crawford (2018), 'Seeing without knowing: limitations of the transparency ideal and its application to algorithmic accountability', *New Media and Society*, **20** (3), 973–89.

Goods, C., A. Veen and T. Barratt (2019), '"Is your gig any good?" Analysing job quality in the Australian platform-based food-delivery sector', *Journal of Industrial Relations*, **61** (4), 502–27.

Jarrahi, M.H. and W. Sutherland (2019), 'Algorithmic management and algorithmic competencies: understanding and appropriating algorithms in gig work', in N. Taylor, C. Christian-Lamb, M. Martin and B. Nardi (eds), *Information in Contemporary Society*, Cham: Springer, pp. 578–89.

Leicht-Deobald, U., T. Busch, C. Schank, A. Weibel, S. Schafheitle, I. Wildhaber, et al. (2019), 'The challenges of algorithm-based HR decision-making for personal integrity', *Journal of Business Ethics*, **160** (2), 377–92.

Meijerink, J.G., A. Keegan and T. Bondarouk (2019), 'Exploring "human resource management without employment" in the gig economy: how online labor platforms manage institutional complexity', paper presented at the Sixth International Workshop on the Sharing Economy, Utrecht, 28–29 June.

Newlands, G. (2020), 'Algorithmic surveillance in the gig economy: the organisation of work through Lefebvrian conceived space', *Organization Studies*, 9 July, 1–19, accessed 12 April 2021 at https://doi.org/10.1177%2F0170840620937900.

Newlands, G., C. Lutz and C. Fieseler (2018), 'Collective action and provider classification in the sharing economy', *New Technology, Work and Employment*, **33** (3), 250–67.

Shapiro, A. (2018), 'Between autonomy and control: strategies of arbitrage in the "on-demand" economy', *New Media and Society*, **20** (8), 2954–71.

Smith, B., C. Goods, T. Barratt and A. Veen (2021), 'Consumer "app-etite" for workers' rights in the Australian "gig" economy', *Journal of Choice Modelling*, **38**, art. 100254, accessed 12 April 2021 at https://doi.org/10.1016/j.jocm.2020.100254.

Van Doorn, N. (2017), 'Platform labor: on the gendered and racialized exploitation of low-income service work in the "on-demand" economy', *Information, Communication & Society*, **20** (6), 898–914.

Wood, A.J., M. Graham, V. Lehdonvirta and I. Hjorth (2019), 'Good gig, bad gig: autonomy and algorithmic control in the global gig economy', *Work, Employment and Society*, **33** (1), 56–75.

Index

'The Platform Economy Puzzles *book is a must-read for every professional with an interest in platform work. It offers a solid overview of the opportunities and challenges of platform-mediated gig work, from both practical and academic perspectives. It is through this integration that the book offers a multisided view on how to realize the benefits of platform work, while limiting its drawbacks.*'
Pim Graafmans, Managing Director YoungOnes, the Netherlands

'*The organization of work in the platform economy is among the key challenges of our age. This is a timely and well-crafted book by leading scholars, covering the topic of the "gig economy" from multiple angles. A must-read for academics and policy-makers alike.*'
Koen Frenken, Utrecht University, the Netherlands

'*The volume brings together an interesting combination of interdisciplinary research. Taking a more critical perspective, some chapters discuss the explosive nature of platform urbanism and data extraction, while others outline the challenges the platform economy poses to labor law. Moreover, the volume directs attention to unique sociotechnical contours of the gig economy such as algorithmic management, multi-party working relationships, and the institutional complexity of human resource management couched in labor platforms. Overall, the book presents helpful implications for various stakeholders, including researchers and practitioners of the gig economy as well as policy-makers.*'
Mohammad Hossein Jarrahi, University of North Carolina at Chapel Hill, USA

'*Masterfully balancing gig work's historic precedents and future potential,* Platform Economy Puzzles *offers a candid and pragmatic introduction to the critical challenges facing society, industry, workers, and policy-makers today. This multi-disciplinary collection by respected scholars not only provokes the reader with critical questions about the state of platform-mediated gig work, but also offers a rare discovery: potential solutions.*'
Gemma Newlands, MSc, PhD candidate, University of Amsterdam, the Netherlands and Doctoral Stipendiary Fellow, BI Norwegian Business School, Norway